A Practical Manual for Transcranial Magnetic Stimulation

Dylan J. Edwards • Peter J. Fried
Paula Davila-Pérez • J. C. Horvath
Alexander Rotenberg • Alvaro Pascual-Leone

A Practical Manual for Transcranial Magnetic Stimulation

Springer

Dylan J. Edwards
Thomas Jefferson University
Moss Rehabilitation Research Institute
Philadelphia, PA, USA

Paula Davila-Pérez
Health Research Institute-Fundacion
Jimenez Diaz University Hospital,
Universidad Autonoma de Madrid
(IIS-FJD, UAM)
Rey Juan Carlos University Hospital
(HURJC), Department of Clinical
Neurophysiology
Madrid, Spain

Alexander Rotenberg
Harvard Medical School
Boston Children's Hospital
Boston, MA, USA

Peter J. Fried
Berenson-Allen Center for Noninvasive
Brain Stimulation
Beth Israel Deaconess Medical Center –
Harvard Medical School,
Boston, MA, USA

J. C. Horvath
LME Global
Scottsdale, AZ, USA

Alvaro Pascual-Leone
Harvard Medical School
Hinda and Arthur Marcus Institute
for Aging Research Hebrew SeniorLife
Boston, MA, USA

ISBN 978-3-031-62303-5 ISBN 978-3-031-62304-2 (eBook)
https://doi.org/10.1007/978-3-031-62304-2

This Springer imprint is published by the registered company Springer Nature Switzerland AG
The registered company address is: Gewerbestrasse 11, 6330 Cham, Switzerland

Preface

Background on Transcranial Magnetic Stimulation and Why This Guide Is Needed

In 1984, the first demonstration of successful transcranial magnetic stimulation (TMS) was performed by Barker and his colleagues (Barker et al. 1985) in an awake, healthy human. This relatively simple procedure, brought about by clever engineering, enabled induction of current in the human brain through the scalp practically painlessly, and with sufficient magnitude to evoke a hand twitch when applied to the contralateral motor cortex. Motor neurophysiologists had a tool like no other: the ability of TMS to painlessly activate the motor cortex in an awake person and produce an observable involuntary muscle twitch (and a measurable motor-evoked potential; MEP) in the hand contralateral to the site of stimulation. This led to a number of insights about central motor conduction pathways and opened the door for a range of studies of human brain function that might previously have required the more uncomfortable and cumbersome technique of electrical stimulation (transcranial or invasive). Clinical neurophysiologists embraced TMS to noninvasively interfere with the activity in the normal brain, activate cortical neurons, and induce observable and measurable phenomena to explore causal relations between brain and behavior.

In the early 1990s, important physiological studies demonstrated that while a single pulse of TMS has subsequent intra-cortical effects lasting <1 s, a train of repetitive pulses at consistent intervals (e.g., 1–20 pulses per second; Hz) could alter local excitability for several tens of minutes. The technique was termed "repetitive transcranial magnetic stimulation" (rTMS). This pivotal work provided evidence that activity could be modulated in specific cortical brain regions, which was appealing as a treatment option for a host of neurological and psychiatric conditions.

Over the past few decades, scientific and clinical interest in TMS has rapidly escalated (Fig. 1). As the field expanded, researchers in academia and industry developed new hardware and software to apply TMS in conjunction with other techniques (e.g., magnetic resonance imaging; MRI, and electroencephalography; EEG) to study the normal, developing, diseased, injured, or aging brain.

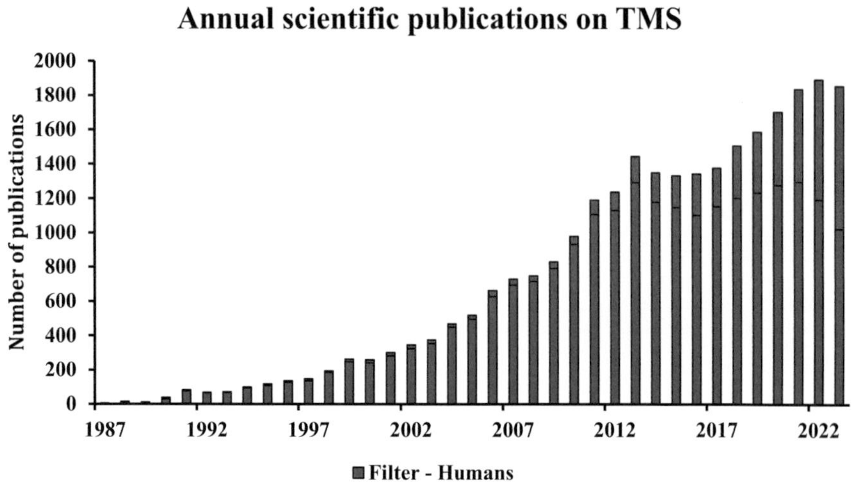

Fig. 1 Scientific interest has escalated since the first demonstration of TMS in 1985, exceeding 1500 publications annually over the past several years (Source: www.ncbi.nlm.nih.gov/pubmed; January, 2024; search terms "Transcranial Magnetic Stimulation," unfiltered, and filtered by "human" studies)

In 2008, the Food and Drug Administration (FDA) in the USA cleared the first TMS device for a therapeutic indication: the Neuronetics' NeuroStar® for the treatment of medication-refractory depression. This was another transformative time for the field, which began to benefit from further industry, investment, public, and academic interest. As of August 2023, seven additional devices have been cleared by the FDA for the treatment of medication-resistant major depressive disorder (MDD) (for a current list, see https://www.clinicaltmssociety.org/tms/devices). In addition, the SAVI Dual™ system from eNeura has been cleared for the treatment of migraine. Most medical insurance companies and Medicare providers across the USA now cover TMS for MDD, and international adoption has similarly continued to grow. New indications are being explored, including several industry-initiated premarket approval studies for TMS devices targeting various nervous system disorders. The increased on- and off-label clinical use is generating a great body of information outside rigid criteria of clinical trials that surely will help the field, along with more traditional research studies, engineering advances, and investigator-initiated studies. Much work is underway, and the outlook is strong for the continued expansion of the field of TMS. In some ways, we are still only at the beginning of the field of TMS-based neurostimulation and neuromodulation, and as the number of practitioners and technicians increases, the need for an easy-access practical manual is growing.

The Intensive Course in Transcranial Magnetic Stimulation at Harvard Medical School

Back in 1998, recognizing the need for formal education and training in TMS, we offered a first continuing medical education (CME) course in TMS. Interest was high and the course was successful, but the feedback from participants was clear: they wanted more hands-on, practical training. This request from students matched our own and our colleagues' needs when faced with the training of new research and clinical fellows, research assistants, and technicians. Thus, we embarked on the development of a curriculum and training course that included theoretical and practical aspects. We cannot overemphasize the contributions of the faculty at the Berenson-Allen Center for Noninvasive Brain Stimulation and the Departments of Neurology and Psychiatry at Beth Israel Deaconess Medical Center and Harvard Medical School. The Education Coordinators at the Berenson-Allen Center [Naomi Pitskel (2004–2006); Andrea Vatulas (2006–2009); Jared Horvath (2009–2011); Sean Kramer (2011–2013); Hillary Reuter (2013–2015); and Alisha Roby (2015–present)] have critically helped evolve the course into its present form: a week-long, intensive immersion into all aspects of TMS in research and clinic, with ample practical, hands-on sessions and a growing flipped-classroom structure complemented by online materials. Most importantly though, the many students have provided invaluable feedback and suggestions and helped us adapt the curriculum to the evolving needs and the growing TMS applications.

In the first two decades, we graduated over 700 participants from 42 countries spanning five continents, and course alumni have incorporated TMS into their clinical practice and research, made important scientific and clinical contributions to the fields of TMS (including over 1000 publications in the peer-reviewed literature), provided ongoing input about the course, and even joined as course faculty.

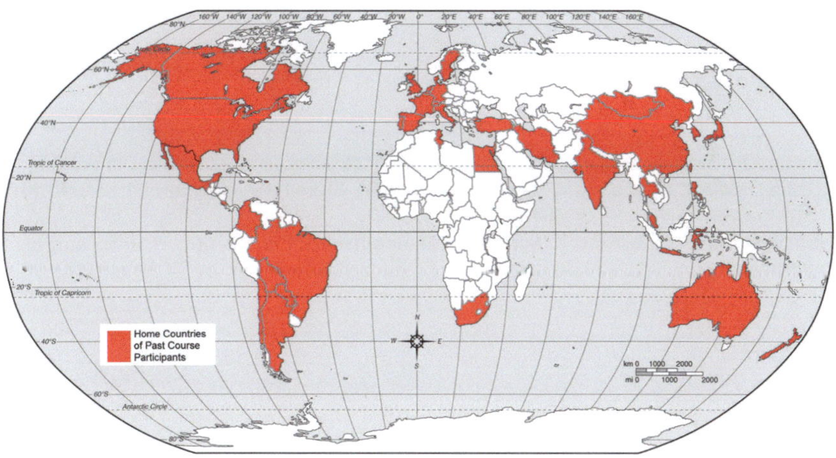

The course continues to attract clinicians, academics, and scientists from a range of fields. Integral to the structure of the course is the broad range of topics covered and applications that include up-to-date TMS research, device engineering, mathematical and preclinical animal models to human research across the lifespan, cognitive and behavioral science applications, and clinical neurological and psychiatric applications. Cutting-edge methods for both understanding the physiologic impact of TMS and refining and optimizing the method are covered with an emphasis on the practical application of TMS, using a range of stimulators and coils. Students in the course receive instruction in small classes as both operators and participants. In addition to basic TMS operations, they are taught technical troubleshooting, as well as experimental and clinical reasoning.

Over the years of the course, with the help and input of the students, we developed a manual that aimed to summarize, in a simple, accessible, and practically useful manner, key concepts and protocols. We recognized that such an abridged version of the various laboratory TMS techniques did not exist, but was often requested by those entering the field, and we therefore decided to develop a practical guide for laboratory and clinical TMS application. This book is not a definitive theoretical account of each TMS technique and does not aim to replace formal training in the use of TMS (Fried, et.al, 2020). Rather, this work is a summary of the most common TMS protocols and an overview of basic TMS principles. We hope that for some it may serve as study guide and for others as a reminder and quick consultation guide.

Philadelphia, PA, USA — Dylan J. Edwards
Boston, MA, USA — Peter J. Fried
Madrid, Spain — Paula Davila-Pérez
Scottsdale, AZ, USA — J. C. Horvath
Boston, MA, USA — Alexander Rotenberg
Boston, MA, USA — Alvaro Pascual-Leone

Further Reading

Barker, AT; Jalinous, R; Freeston, IL (1985). "Non-Invasive Magnetic Stimulation of Human Motor Cortex". *The Lancet* **325** (8437): 1106–1107. doi:https://doi.org/10.1016/S0140-6736(85)92413-4. PMID 2860322.

Fried PJ, Santarnecchi E, Antal A, Bartres-Faz D, Bestmann S, Carpenter LL, et al. Training in the practice of noninvasive brain stimulation: Recommendations from an IFCN committee. Clin Neurophysiol. 2020 Dec 3;

With Contributions by

Lorella Battelli, Ph.D.
Joan Camprondon, M.D., Ph.D.
Bernard Chang, M.D.
Roch M. Comeau, Ph.D.
Ann Connor, R.N.
Mar Cortes, M.D.
Faranak Farzan, Ph.D.
Shirley Fecteau, Ph.D.
Joyce Gomes-Osman, P.T., Ph.D.
Mark Halko, Ph.D.
Ali Jannati, M.D., Ph.D.
Jari Karhu, Ph.D.
Umer Najib, M.D.
Lindsay Oberman, Ph.D.
Juha Silvanto, Ph.D.
Marine Vernet, Ph.D.
Anne Weigand, Ph.D.

Contents

About the Authors

Paula Davila-Perez (MD, PhD) is a clinical neurophysiologist and a noninvasive brain stimulation researcher. Paula has extensive experience in human clinical neurophysiology including electromyography, electroencephalography and evoked potentials. Her investigations focus on brain plasticity, the influence of electromagnetic fields in neural networks and brain excitability. Dr. Davila-Perez is interested in better understanding brain mechanisms underlying cortical plasticity changes and motor and cognitive deterioration. These investigations allow her to improve the diagnosis and treatment of different neuropsychiatric diseases. Dr. Davila-Perez was involved in the research and clinical transcranial magnetic stimulation (TMS) programs of Berenson-Allen Center for Noninvasive Brain Stimulation and lectured in the Continuing Medical Education course on TMS at Harvard Medical School for 7 years.

Dylan J. Edwards is the Nancy Wachtel Shrier Director of Moss Rehabilitation Research Institute in Philadelphia, USA, and Professor in the Sidney Kimmel Medical College at Thomas Jefferson University. He has been conducting research in human neurophysiology and noninvasive neurostimulation since 1994. His research focuses on translating basic science findings in neurologically healthy people to develop and evaluate novel, clinically-relevant approaches for neurological assessment and neurorehabilitation in stroke and spinal cord injury. Dr. Edwards has published numerous peer-reviewed papers and given many invited lectures across the globe on noninvasive brain stimulation and transcranial magnetic stimulation (TMS). In addition, he taught and co-led a Continuing Medical Education course on transcranial magnetic stimulation at Harvard Medical School (Berenson-Allen Center for Non-Invasive Brain Stimulation) for 10 years, and co-developed the published technical training standards for noninvasive brain stimulation for the International Federation for Clinical Neurophysiology. Dr. Edwards is currently leading translational research studies to advance precision targeting using contemporary TMS brain mapping methods and to test novel TMS treatment approaches in neurorecovery.

Peter J. Fried is Associate Director for Research of the Berenson-Allen Center for Noninvasive Brain Stimulation at Beth Israel Deaconess Medical Center in Boston, USA, and Assistant Professor of Neurology at Harvard Medical School. Dr. Fried received his Ph.D. in Anatomy and Neurobiology from the Boston University School of Medicine, and he has been conducting research in human neurophysiology and noninvasive neurostimulation since 2009. His research focuses on using neuroimaging and noninvasive brain stimulation to investigate the neurocognitive changes in healthy and disordered aging. Dr. Fried has published over 40 peer-reviewed papers. In addition, he taught a Continuing Medical Education course on transcranial magnetic stimulation at Harvard Medical School (Berenson-Allen Center for Non-Invasive Brain Stimulation) for 10 years and co-developed the published technical training standards for noninvasive brain stimulation for the International Federation for Clinical Neurophysiology.

J. C. Horvath (PhD, MEd) is an educational neuroscientist with expertise in human learning, memory, and attention. He has conducted research and lectured at Harvard University, The University of Melbourne, Monash University, and over 750 schools internationally. Jared has published 6 books, over 50 research articles, and his work has been featured in numerous popular publications, including *The New Yorker*, *The Atlantic*, *The Economist*, and ABC's *Catalyst*. He currently serves as Director of LME Global: a team dedicated to bringing the latest brain and behavioural research to teachers, students, and parents alike.

Alvaro Pascual-Leone is Medical Director of the Deanna and Sidney Wolk Center for Memory Health and a Senior Scientist at the Hinda and Arthur Marcus Institute for Aging Research at Hebrew SeniorLife. He is also a Professor of Neurology at Harvard Medical School. His research focuses on cortical plasticity, neuromodulation, diagnostic and therapeutic applications of noninvasive brain stimulation for neuropsychiatric conditions, and maintenance of brain health across the lifespan. Dr. Pascual-Leone has been a pioneer in the field of transcranial magnetic stimulation (TMS) since its inception. As one of the world's leading experts in the field, he has published prolifically and presented widely on TMS and noninvasive brain stimulation. He founded the Berenson-Allen Center for Noninvasive Brain Stimulation at the Beth Israel Deaconess Medical Center and Harvard Medical School and served as Director there for over 20 years. He also developed and directed the first Continuing Medical Education training course on TMS at the Berenson-Allen Center for Noninvasive Brain Stimulation. The work of Dr. Pascual-Leone spans pre-clinical and clinical science, clinical care, technical development, education, and policy.

Alexander Rotenberg is Professor of Neurology at Harvard Medical School, Director of the Neuromodulation Program at Boston Children's Hospital, and Director of the Experimental Neurophysiology Core at the Kirby Center for Neurobiology. In his research, Dr. Rotenberg is developing methods to treat brain injury that can lead to epilepsy, by identifying and manipulating biological targets using noninvasive stimulation and novel drugs. He leads a basic science laboratory researching applications of noninvasive brain stimulation for models of epilepsy and traumatic brain injury. In addition, Dr. Rotenberg is conducting pre-clinical studies and clinical trials to better understand the neurobiology of brain stimulation and test potential clinical therapies. He is also leading efforts to adapt methods for noninvasive brain stimulation, including transcranial magnetic stimulation (TMS) to pediatric populations with epilepsy and other neurological conditions. Dr. Rotenberg is Co-Director of the Continuing Medical Education training course on TMS at the Berenson-Allen Center for Noninvasive Brain Stimulation, and he has made substantial contributions to the field, both in animal models of neurological conditions and clinical populations.

List of Abbreviations

%MSO	Percent of maximum stimulator output
ADM	Abductor digiti minimi
AMT	Active motor threshold
AP	Anterior-to-posterior
APB	Abductor pollicis brevis
AP-PA	Anterior-to-posterior – Posterior-to-anterior
C7	Seventh cervical vertebra
Ca	Calcium
CBI	Cerebellar-brain inhibition
CMAP	Compound motor action potential
CMCT	Central motor conduction time
CME	Continuing Medical Education
CNS	Central nervous system
CoG	Center of gravity
CS	Conditioning stimulus
cSP	Contralateral cortical silent period
cTBS	Continuous theta burst stimulation
DLPFC	Dorsolateral prefrontal cortex
D-wave	Direct wave
EEG	Electroencephalography
E-field	Electric field
EMG	electromyography
FDA	Food and Drug Administration
FDI	Frist dorsal interosseous
fMRI	Functional magnetic resonance imaging
GABA	γ-aminobutyric acid
I/O curve	Input/Output curve
ICF	Intracortical facilitation
IFCN	International Federation of Clinical Neurophysiology
IHI	Interhemispheric inhibition
IRB	Institutional Review Board

ISI	Inter-stimulus interval
iSP	Ipsilateral cortical silent period
iTBS	Intermittent theta burst stimulation
I-waves	Indirect waves
L1	First lumbar vertebra
LICI	Long-interval intracortical inhibition
LM	Lateral-medial
LTD	Long-term depression
LTP	Long-term potentiation
M1	Primary motor cortex
MDD	Major depressive disorder
MEP	Motor evoked potential
MRI	Magnetic resonance imaging
MT	Motor threshold
Na	Sodium
NMDA	N-methyl-D-aspartate
PA	Posterior-to-anterior
PAS	Paired-associative stimulation
PCT	Peripheral conduction time
PET	Positron emission tomography
PNS	Peripheral nervous system
RMT	Resting motor threshold
rTMS	Repetitive transcranial magnetic stimulation
SAI	Short afferent inhibition
SAS	Spinal associative stimulation
SICI	Short-interval intracortical inhibition
SP	Cortical silent period
TBS	Theta-burst stimulation
TCT	Total conduction time
TMS	Transcranial magnetic stimulation
TS	Test stimulus
TST	Triple-pulse stimulation technique
V1	Primary visual cortex
V2	Secondary visual cortex
V5/MT+	visual area 5/middle temporal visual processing area

Chapter 1
Basic Mechanisms

Abstract The transcranial magnetic stimulation (TMS) device, together with the coil held adjacent to the scalp in the awake human, can painlessly generate electric fields in brain tissue via electromagnetic induction. The rate of discharge and pulse shape of the capacitor, the magnitude of discharge, coil structural characteristics, and the orientation and distance from the coil to brain tissue fundamentally alter the resultant brain electric field. Variation exists in the strength, depth, and area and current direction of the brain electric field, according to characteristics of the specific device, coil, and parameter settings. Several device types exist with different purposes and parameter settings to match the scientific and clinical applications.

D. J. Edwards et al., *A Practical Manual for Transcranial Magnetic Stimulation*, https://doi.org/10.1007/978-3-031-62304-2_1

1.1 How TMS Works, Part 1: *General Principles of Electromagnetic Induction*

Transcranial magnetic stimulation (TMS) is electrode-less electric stimulation of brain tissue via *electromagnetic induction*. Electromagnetic induction was discovered in 1831 by Michael Faraday who described five key principles that are relevant to proper TMS understanding and practice (below). To fully understand these principles, it is important to keep in mind that electric currents and magnetic fields are linked phenomena with a perpendicular orientational relationship.

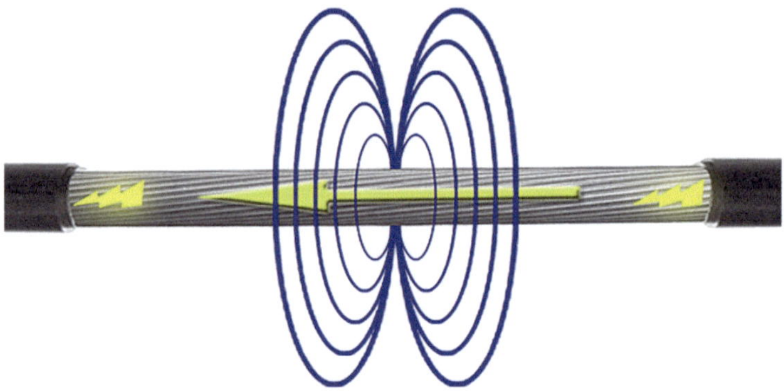

Principle 1: Rate of Change

Faraday discovered that a pulse of electricity could be generated along a metallic coil simply by moving a magnet through it. Faraday observed that the quicker the magnet moved through the coil, the stronger the induced current. Therefore, the *rate of change* of a magnetic field (not simply its strength) determines the strength of the induced current in a nearby conductor. Faraday determined that not only will an electric current generate a magnetic field, but also a fluctuating magnetic field can generate an electric current in a nearby conductor.

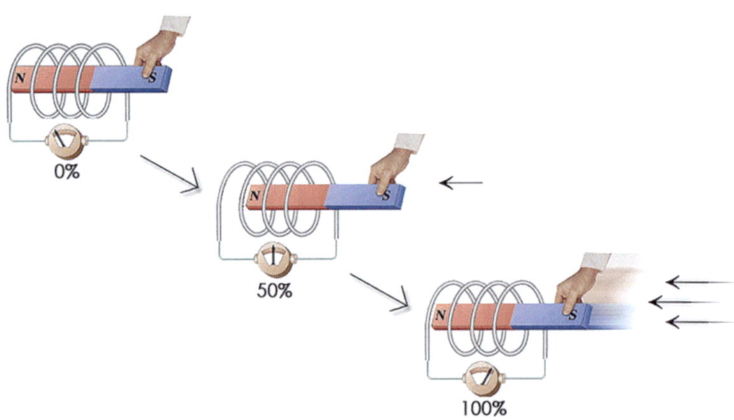

Principle 2: Intensity

Faraday showed that if an electric current was *pulsed* through a coil of wire, a secondary current would be induced in a nearby conductor. In addition, Faraday found that the strength of the secondary current was influenced by the strength of the initial (source) current. Interestingly, Faraday noted that a secondary current would *not* be generated if the initial current was held constant (rather than pulsed). When combined with principle 1 (*rate of change*), Faraday determined that an initial, fluctuating electric current would produce a similarly fluctuating magnetic field which, in turn, would induce the secondary current. As shown in the figure below, the intensity of the induced current increases with the intensity of the source current.

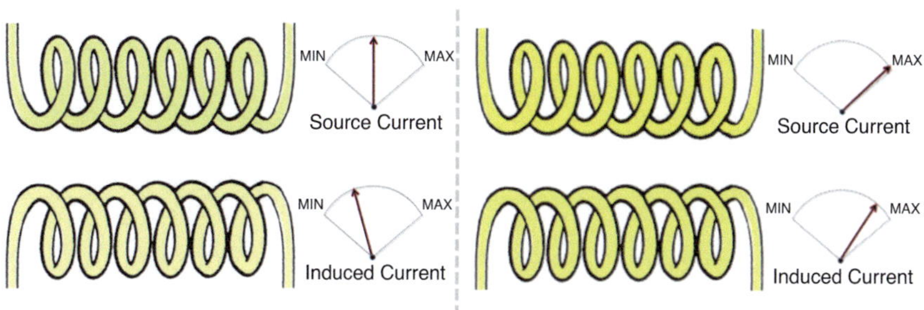

Principle 3: Current Direction

Faraday showed that the secondary induced current flowed in the opposite direction of the initial, source current.

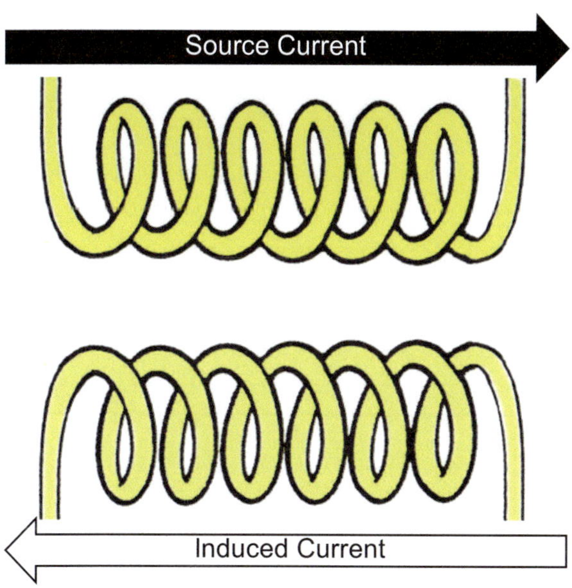

Principle 4: Distance

Faraday showed that electromagnetic induction is strongly influenced by the distance between two interacting coils. The power of a magnetic field decays exponentially with distance from its source. Therefore, as the distance between the two interacting coils increases, the strength of the induced current will decrease as well. This means that the current induced by TMS will always be greatest close to the stimulation coil and decay steeply with distance.

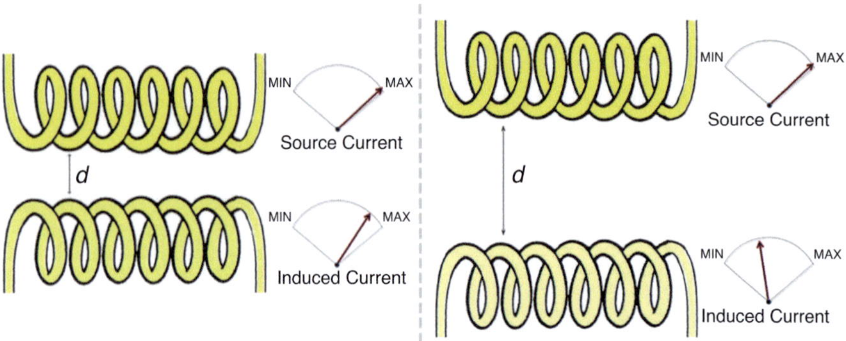

Principle 5: Coil Orientation

Faraday showed that electromagnetic induction was maximal when two interacting coils were aligned in parallel. Conversely, no induction occurred when the two coils were aligned perpendicular. Thus, the amount of induced current will decrease as the orientation of the conductor coil rotates from parallel to perpendicular.

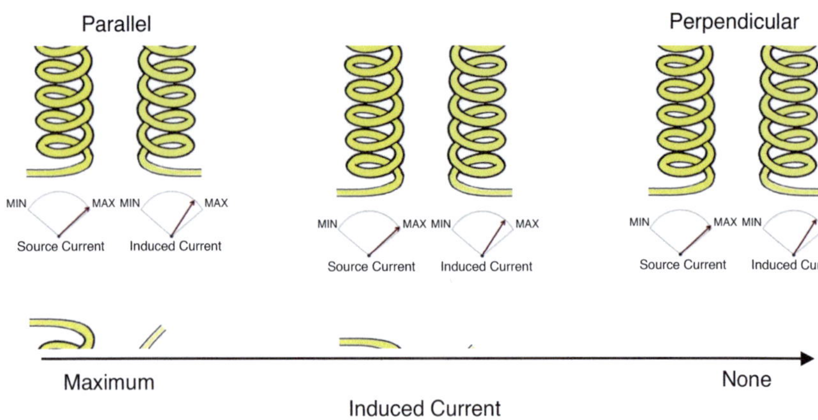

1.2 How TMS Works, Part 2: *Electromagnetic Induction and the Brain*

The basic principles of electromagnetic induction (*Sect. 1.1*) illustrate how TMS can stimulate the brain without the use of electrodes. When a strong fluctuating current is passed through a stimulating coil, it generates an equally fluctuating magnetic field that emanates perpendicular to the coil surface. When the stimulating coil is held against an individual's head, this magnetic field passes through the scalp, skull, and meninges. When the magnetic field reaches the adjacent cortex, a secondary current is induced with neural elements acting as conductors.

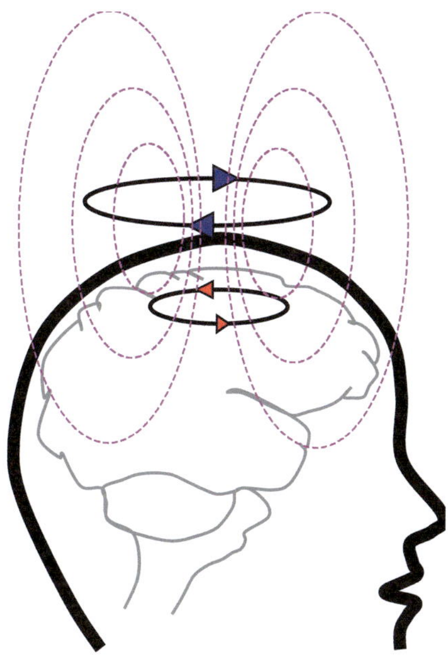

As a fast *rate of change* (see *Sect. 1.1*) of the magnetic field is critical to inducing a secondary current, the initial electric pulse through the TMS coil is very brief with a steep ramp on/off.

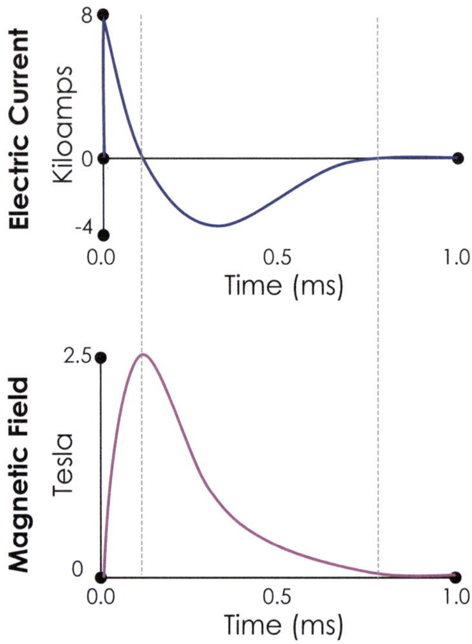

Each of Faraday's principles applies to TMS. As shown in the figure below, the strength of the initial current, the distance of the coil from the cortex, the orientation of the neurons (or more specifically, neural elements) to the coil, and the direction of the initial current will all impact the characteristics of the secondary current induced in the cortex. Additionally, the cytoarchitecture of the underlying cortical column will also impact the characteristics of this induced current. As such, each of these variables is important to keep in mind and manipulate, if possible, when determining optimal TMS procedures.

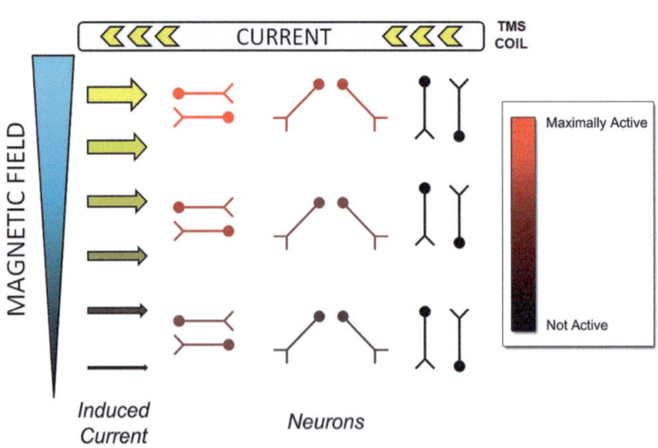

1.3 The TMS Device

The design and components of TMS devices are relatively straightforward and universal. Each machine will have a "main unit" and a stimulating coil.

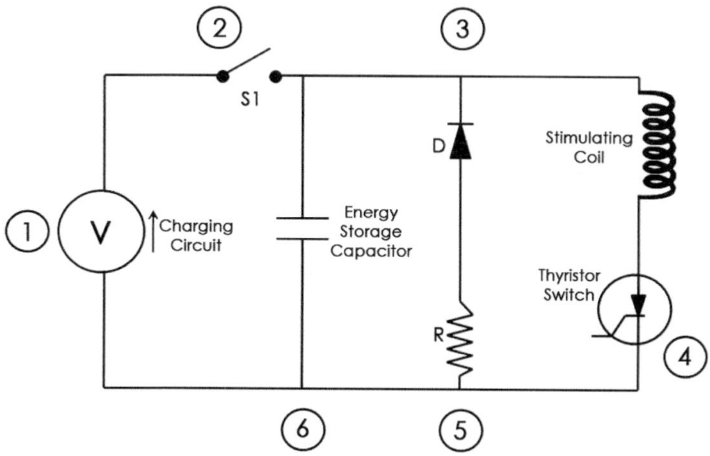

The main unit is composed of several important components:

1. *Charging system*—this system generates the current used to generate the magnetic field essential to TMS. A typical charging system can quickly generate 500 J.
2. *A discharge switch*—this will take the energy from the storage capacitor. Occasionally this switch will be hooked to a pneumatic foot-pedal trigger or an onboard button-press trigger (varies between machines/manufacturers).
3. *Energy recovery circuitry*—to recharge the unit.
4. *Thyristor*—Thyristors are electrical devices capable of switching large currents over a short period of time. In this case, the thyristor acts as the bridge between the capacitor and coil, transferring 500 J between the two in less than 100 ms.
5. *Pulse-shape circuitry*—can generate monophasic or biphasic pulses (*Sect. 1.5*).
6. *One or more energy storage capacitors*—these allow for multiple pulses to be stored and generated in quick succession. Multiple storage capacitors are required for rTMS, theta-burst stimulation (TBS), and any other non-single pulse paradigm.

Stimulating Coil

The stimulating coil is the only part of the machine to come into contact with the subject/patient. Varied coil types can deliver varied pulse shapes and sizes from the device (*Sect. 1.4*).

Examples of Commercially Available Units

Several commercially available devices are depicted below. Although this is not a comprehensive look at devices, it provides a sense of the universal components required for TMS machines.

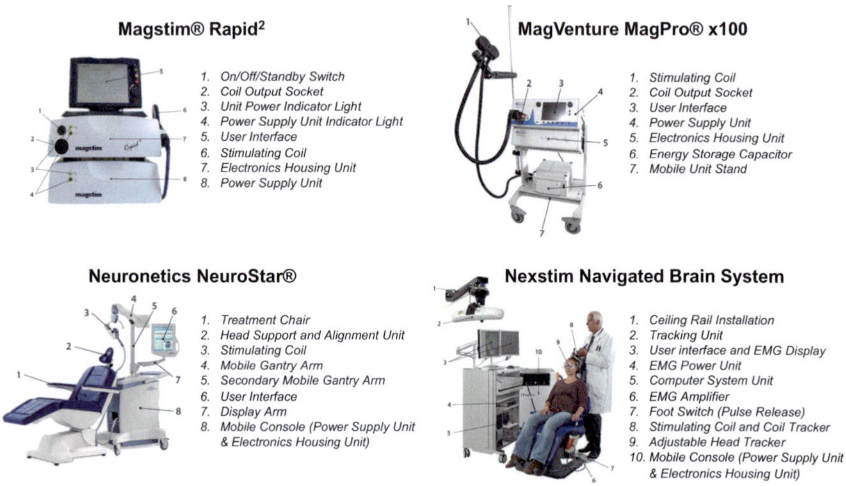

1.4 The TMS Coil

TMS coils are typically made of copper wire encased within an insulating material (typically hard plastic). Accordingly, no direct electrical contact is ever made between the coil and the participant's head. Coils can be wound tightly or loosely, and the wire can be situated either close or far from the surface of the insulating casing.

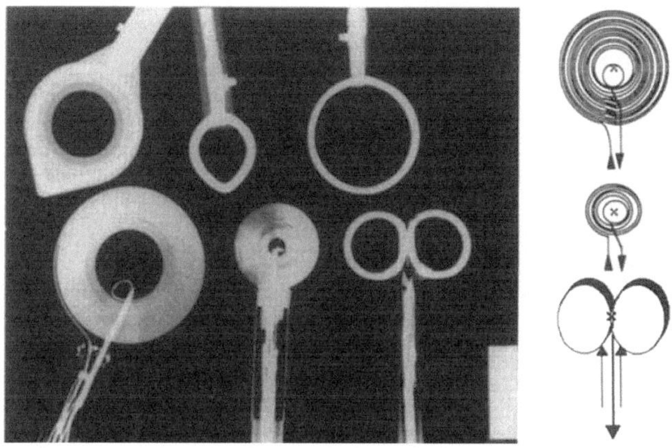

Due to the energy transfer from the electric current, coils will heat up with repeated usage, which will occur faster at higher intensities. Most TMS systems have temperature sensors imbedded in the coil that disables the triggering mechanism if the temperature approached an unsafe temperature (~40 °C) to prevent burning the participant or damaging the components. There are several solutions to dissipate this heat and allow for rapid/continual stimulation at high intensities. Some coils have an iron core, which slows the heating time of the coil but increases the weight considerably. An alternative is a cooling system that circulates either a special heat-transfer fluid or air. Coils can also have static/non-circulating fluid.

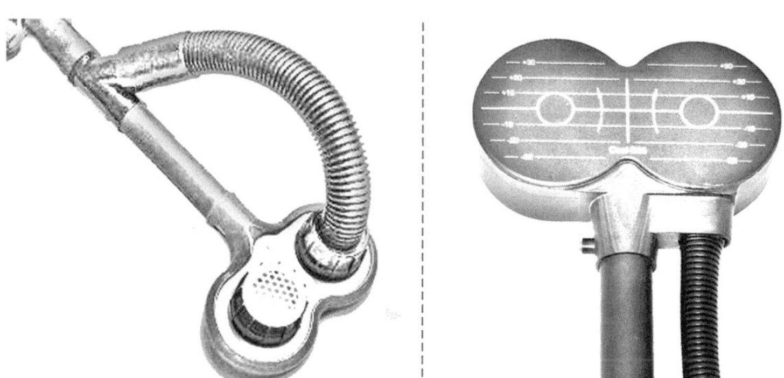

Coils are available in a variety of shapes and sizes. The geometry of the coil determines the shape, depth, and focality of the induced current. The optimal coil, therefore, is dependent upon the unique protocol. Most commercially available coils utilize the following designs:

Round Coil

The round coil was the first TMS coil design. The magnetic field generated by this type of coil is maximal at the edges and zero at the center.

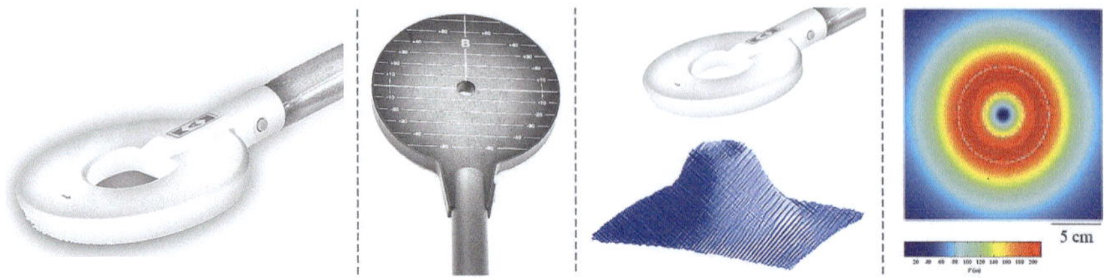

Figure-of-Eight Coil

Easily the most recognizable and utilized of coil designs, the figure-of-eight is formed by abutting two single round coils against one another. Although the pattern from each round circular coil may be relatively non-focal, the *combined* magnetic field (summed at the contact point between coils) is stronger than in surrounding regions and is relatively easy to determine spatially. Although unproven, mathematical modeling suggests a small figure-of-eight (each "wing" measuring 4 cm in diameter) can achieve a spatial resolution of approximately 5 mm^3 of brain volume. This type of coil is preferred for most clinical and academic utilizations of TMS—including repetitive and chronometric measures.

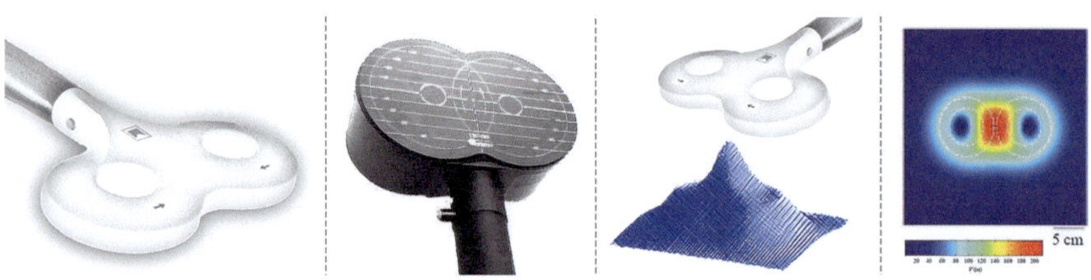

H-Coils (Deep TMS)
H-Coils aim to stimulate deeper, non-superficial cortical layers. This is achieved through the proper placement and interaction of multiple electromagnetic fields around the head. Typically encased in a large helmet, H-coils have been demonstrated to stimulate neural regions up to a depth of 6 cm (compared to 1.5–2.5 cm depth for a common figure-of-eight coil). It is important to note, however, that deep stimulation is not "selective"—meaning current is also induced in the brain volume superficial to the intended "deep" target.

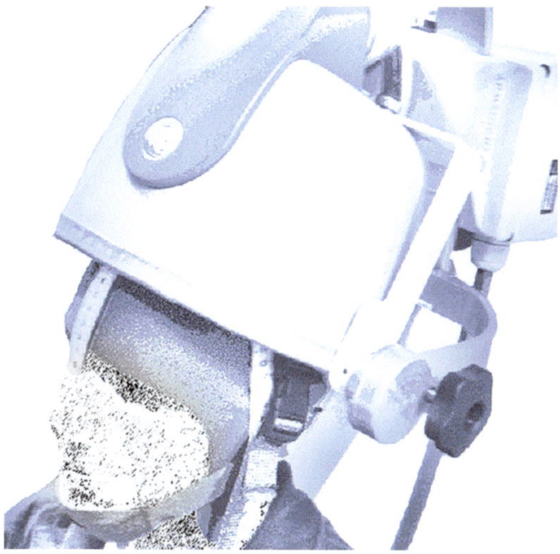

Additional Coil Types
Coils can be made to specification. As such, several experimental coil designs have been created, including the double coil, the horizontal racetrack, double-cone (angled) coil, the vertical racetrack, and small-animal sized coils.

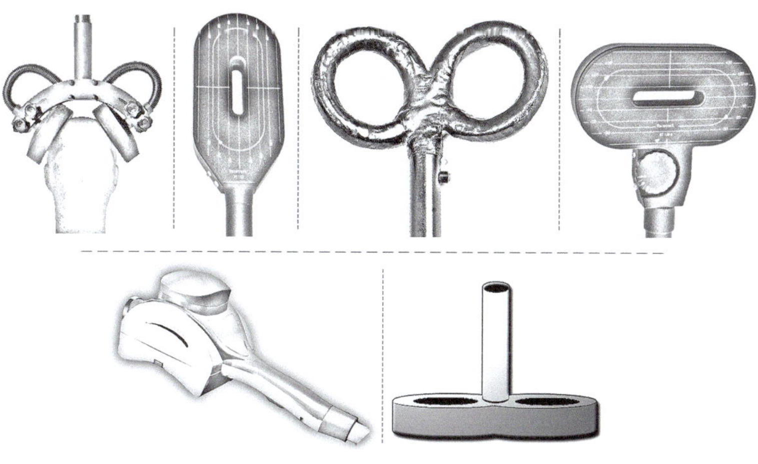

1.5 Characteristics of the Electromagnetic Pulse

Waveform

Monophasic pulses generate a flow of current primarily in one direction with a corresponding induced current in the opposite direction within any nearby conductor. In TMS, monophasic pulses are often used to selectively activate specific neural elements or brain structures that differentially respond to particular current directions depending on the intensity of stimulation. For example, in the primary motor cortex, a lateral-medial (LM) pulse will typically elicit direct activation of the layer-V pyramidal neurons (i.e., a direct [D]-wave descending cortico-spinal volley), while pulses oriented posterior-anterior (PA) and anterior-posterior (AP) have been shown to indirectly activate layer-V pyramidal neurons resulting in early and late cortico-spinal volleys (indirect [I]-waves), respectively. In addition, different orientations of the induced current have been shown to require different intensities of stimulation (i.e., thresholds) to elicit specific descending cortico-spinal volleys, with PA showing the lowest threshold and LM having the highest (see Further Reading for more comprehensive information on this topic). Lastly, it has been suggested that the electric field (E-field) induced in the cortex has a separate spatial distribution depending on the pulse orientation. Understanding how pulse shape and orientation interact with different neural elements is necessary to interpret the results of cortical mapping and other TMS-based neurophysiological assessments.

Biphasic pulses generate opposing current oscillations, although the individual phases of the oscillation may not be symmetrical. Like monophasic pulses, the order of each phase may be reversed (i.e., AP-PA versus PA-AP). The impact of directionality for biphasic pulses is less well understood and depends on intensity; however, the end result will likely be a mixture of both AP and PA phases.

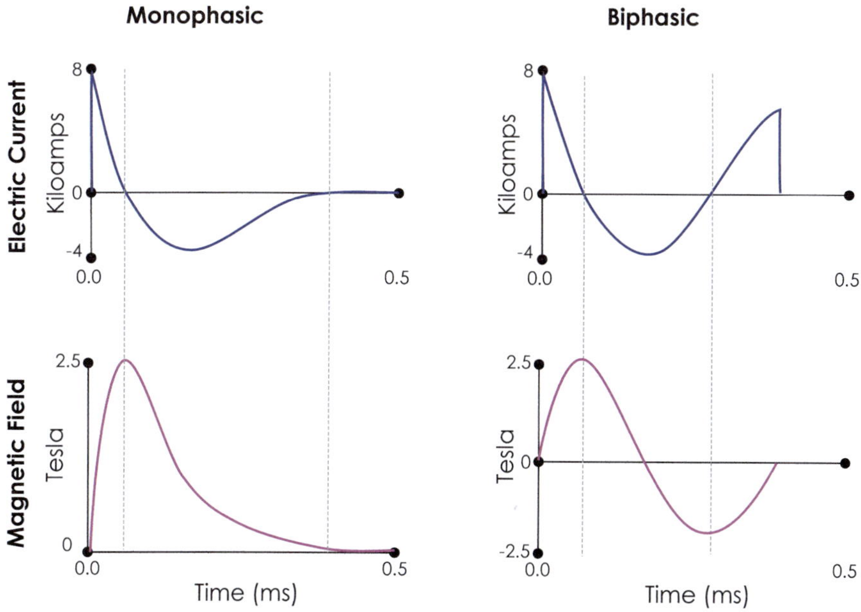

Pulse Intensity

The amount of current passed through the coil can be adjusted. The intensity of the induced current depends on that of the current passed through the coil as well as the rate of change of the magnetic field (*Sect. 1.1*). Each TMS device (*Sect. 1.3*) displays pulse intensity as a percent of the maximum stimulator output (%MSO; represented by green numbers in the figure below). As %MSO is an arbitrary unit, it is necessary to relate the stimulation intensity to a measure of cortical reactivity such as the individual's motor threshold (*Sect. 3.4*) to compare stimulation intensities across devices, coils, and individuals as well as to adhere to safety guidelines (*Sect. 2.2*).

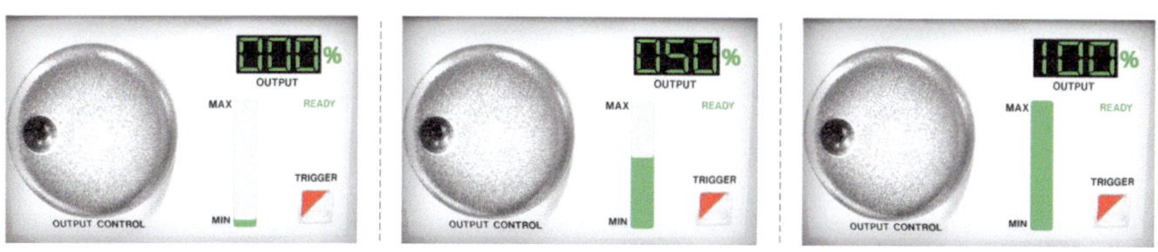

Further Reading

Di Lazzaro V. & Rothwell J.C. (2014). Corticospinal activity evoked and modulated by non-invasive stimulation of the intact human motor cortex. *Journal of Physiology*, 592(19), 4115-4128.

Sommer, M., Alfaro, A., Rummel, M., Speck, S., Lang, N., Tings, T., & Paulus, W. (2006). Half sine, monophasic and biphasic transcranial magnetic stimulation of the human motor cortex. *Clinical Neurophysiology*, 117(4), 838-844.

Chapter 2
Safety

Abstract The application of human transcranial magnetic stimulation should be with due consideration of potential safety concerns and risk for the participant and TMS administrator. Here we indicate general safety issues and potential risks relevant to common TMS settings, as a brief reminder or introduction to key elements, for a laboratory or desk reference. Operation of the TMS device should only occur with appropriate training, pre-reading of published safety recommendations by the International Federation of Clinical Neurophysiology (IFCN), and local university or clinical approvals, such as an Institutional Review Board (IRB) or Medical Executive Committee.

D. J. Edwards et al., *A Practical Manual for Transcranial Magnetic Stimulation*,
https://doi.org/10.1007/978-3-031-62304-2_2

2.1 Health and Safety Concerns

General Safety Concerns

1. Scalp Electrode Heating

When combining TMS with electroencephalography (EEG), induced eddy currents within the EEG electrode can heat and potentially burn the scalp. To avoid this, it is recommended to use electrodes made from low-conductivity plastic or ones that have a radial notch.

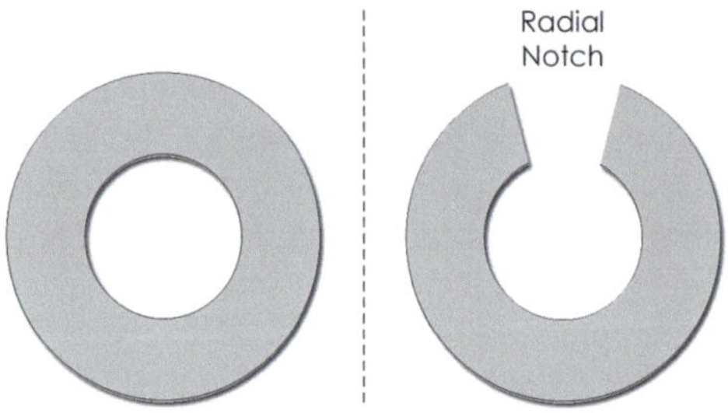

2. Magnetization

As described in Sect. 1.2, TMS involves the generation of a strong electromagnetic field. Consequently, any external or implanted ferromagnetic medical device has the potential to move from interaction with the electromagnetic field. Only those devices cleared for use with high magnetic fields (including TMS and magnetic resonance imaging; MRI) should be used. Furthermore, all conductive material or electronics near the head (hearing aids, hair pins/clips, earphones, etc.) should be removed prior to stimulation. Jewelry may need to be removed if it is directly under the coil.

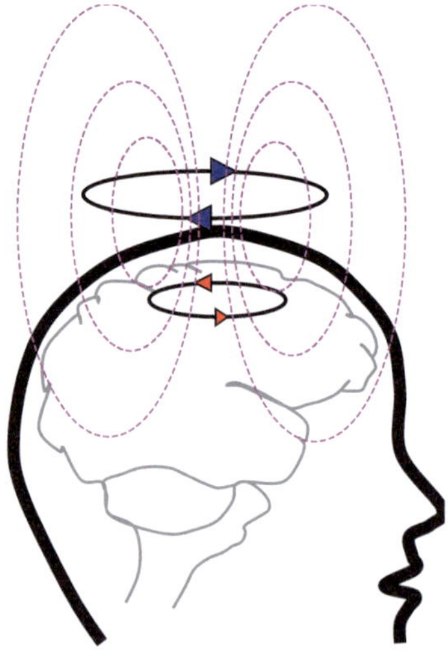

3. Induced Currents

Any wires or electronic devices near an active TMS coil may be affected by the electromagnetic field causing damage. For example, EEG electrode wires (or similar) should not be looped and instead bound in a twisted fashion.

Electrical currents can also be induced in the circuitry of implanted medical devices—whether switched ON or OFF. As described above, only approved devices should be used, and any wires should be free of loops.

4. Implanted Electrodes

Although TMS appears to be safe for patients with implanted stimulators, provided the system is not within close proximity to the TMS coil, specific safety parameters (such as distance from the coil) have not yet been determined. Accordingly, extra care should be taken in these participants and TMS should only be performed after careful evaluation and when there is a medical justification (e.g., depression with suicidal ideation).

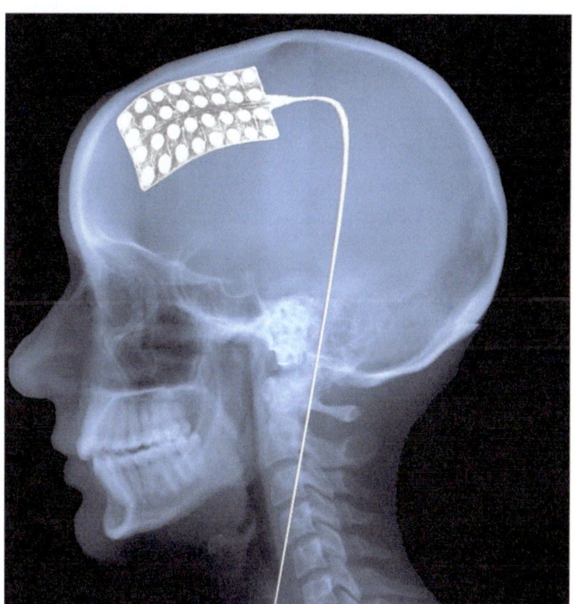

Side Effects

Common

1. Headache/Neck Pain

Approximately 20–40% of subjects undergoing TMS have reported headache and/or neck pain. While transient headache or muscle tension can occur directly from TMS-induced contraction of superficial muscles (especially over frontal or occipital locations and around the sensitive muscles of the orbit), it is more common that these sensations result from a tight-fitting swim cap or headband, excessive pressure of the TMS coil on the subject's head, or the participant's posture during stimulation. These can be minimized by proper coil handling and monitoring feedback from the participant. In most cases, any headache or muscle tension resolves on its own; however, a single dose of over-the-counter pain medication can help alleviate this pain if it persists beyond stimulation.

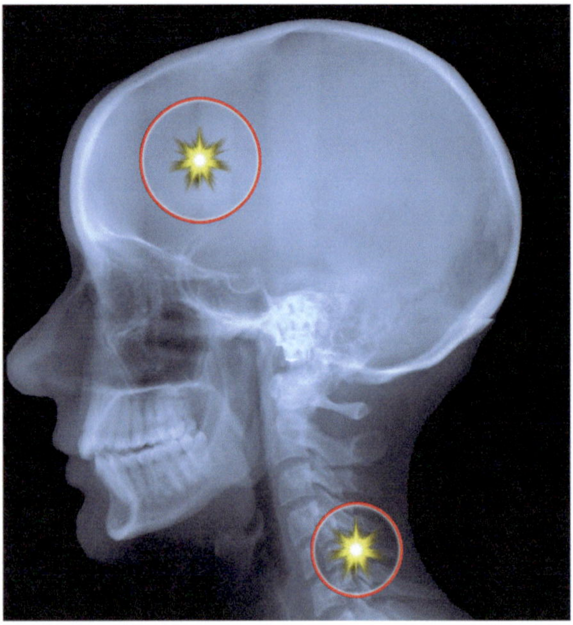

In daily clinical repetitive TMS (rTMS) treatments, local pain generated during prefrontal stimulation has been reported to decrease over the first 3 days of treatment. Accordingly, the intensity of stimulation may be slowly increased over the first few days of treatment.

To our knowledge, no migraines have been reported to date following TMS.

2. Acoustic Trauma

The TMS pulse generates a compression air wave, which at high intensities can reach acoustic levels as high as 120–140 dB that can damage hearing. TMS performed without adequate hearing protection can lead to tinnitus and acute or permanent hearing loss. Therefore, all individuals (including participants, technicians, and visitors/family members) are strongly recommended to wear adequate hearing protection (e.g., earplugs) at all times during stimulation. There may also be requirements from federal, local, or institutional occupational health and safety boards.

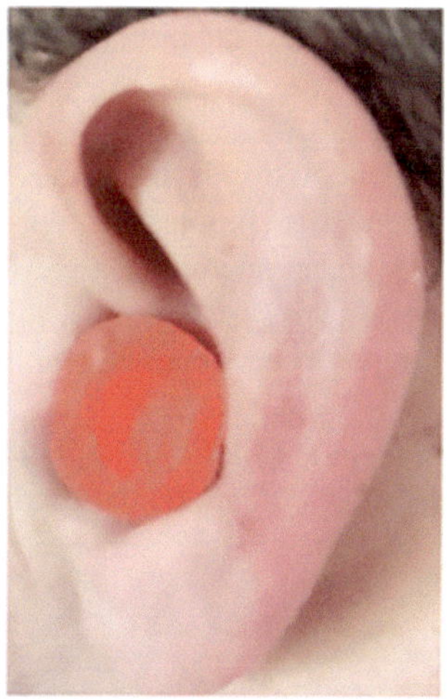

Rare

1. Seizure

TMS-induced seizures are a major concern and, although exceedingly rare, have been reported in healthy subjects and patient populations. It is important to establish an emergency seizure-response plan prior to any stimulation protocol (*Sect. 2.3*).

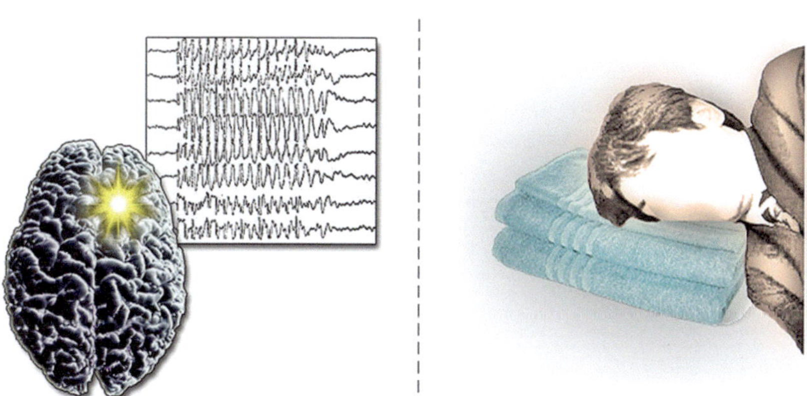

2. Syncope/Fainting

Syncope and pre-syncope symptoms can occur in the context of TMS and may appear similar to a seizure. While the precise cause of syncope is multifactorial (including the stimulation itself, anxiety, physical or psychological distress, or situational factors such as room temperature, dehydration, etc.). It is important to be able to recognize the symptoms of pre-syncope and distinguish syncope from seizure (*Sect. 2.3*). Prior to stimulation, all participants should be encouraged to immediately report any symptoms of pre-syncope (e.g., feeling dizzy, lightheaded, or nauseated). In the case of syncope/pre-syncope, all stimulation should be ceased, assistance offered, and a neurological evaluation undertaken if appropriate.

3. Mood

There is a chance that rTMS could induce changes in mood, such as an acute manic state (especially in patients with bipolar disorder, although the prevalence of this is thought to be below the typical rate of disease cycling).

Transient psychotic symptoms, anxiety, insomnia, suicidal ideation, and extreme agitation following rTMS have been reported (although their prevalence is extremely low).

In the instance of extreme mood shifts, stimulation should be immediately ceased, assistance offered, and a neurological and/or psychiatric evaluation administered.

4. Dental Pain

Several instances of induced dental pain have been reported. If this occurs, it may indicate a dental issue. Stimulation should be stopped, and the participant should be encouraged to seek a dental evaluation.

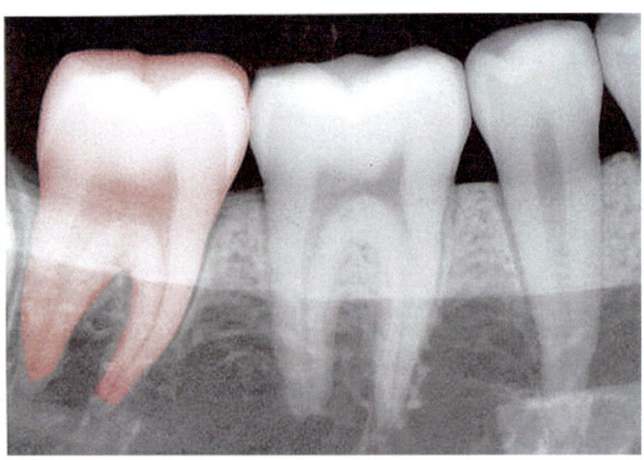

Contraindications

It is important to follow all published guidelines on TMS contraindications. Although stimulation decisions should be made on a case-by-case basis, deviations from guidelines should be made only for a medically compelling reason and taking the risk/benefit ratio into account.

1. Implanted Cranial Electrodes/Metallic Hardware

Due to the risks of heating, displacement, and electromagnetic interference, the presence of implanted electrodes or metallic hardware near the site of stimulation poses a relative contraindication for TMS. Electrodes may be more of a risk than thin titanium skull plates, which interact minimally with the TMS-induced electromagnetic field and are secured to the skull.

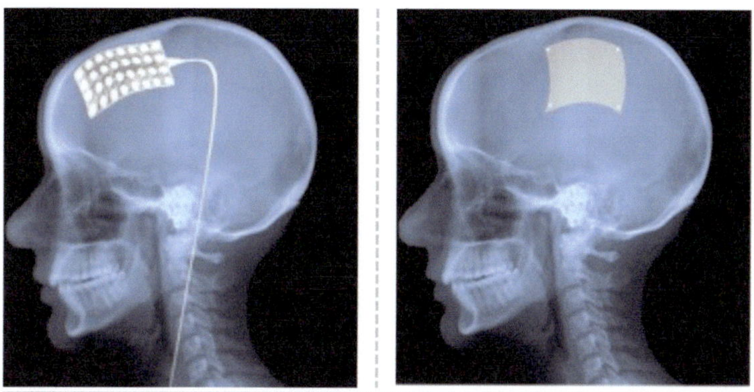

2. Cochlear Implants

In most cases, it is not recommended to administer TMS to an individual with a cochlear implant due to the risks of heating, displacement, and electromagnetic interference.

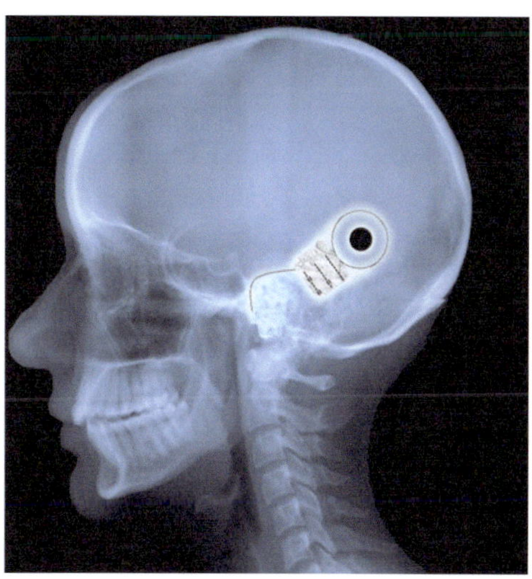

3. Personal History of Seizure, Epilepsy, or Lowered Seizure Threshold

A history of seizure, personal or immediate family history of epilepsy (recurrent seizures), and drugs and medical conditions that lower seizure threshold (e.g., stroke, brain tumor, sleep deprivation, cortical malformation) may put an individual at higher risk for a TMS-induced seizure (*Sect. 2.3*). Single-pulse TMS has been used for motor mapping in adults and children with epilepsy and appears to be minimal risk; however, even single pulse near a cortical area at risk may be sufficient to induce a seizure. Applying rTMS, particularly high-frequency rTMS, carries the highest risk of inducing seizures, and should be used with caution in individuals at risk and a seizure identification and response plan should be put into place (*Sect. 2.3*).

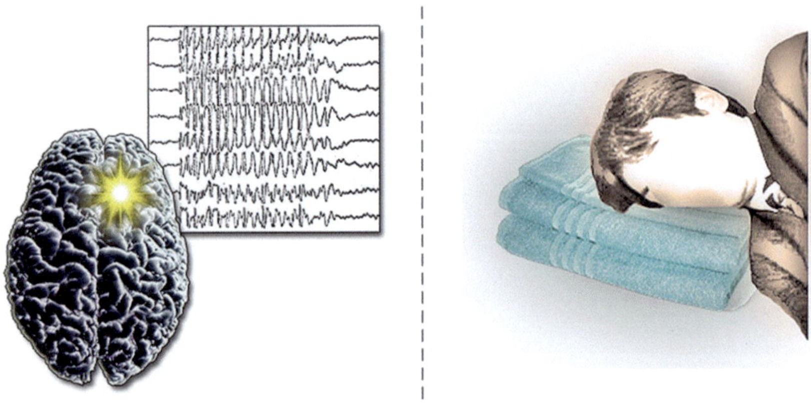

4. Cerebral Lesion

Due to current shunting and risk of seizure, TMS should be used cautiously in individuals with vascular, traumatic, tumoral, infectious, or metabolic lesions with appropriate oversight, such as an Institutional Review Board (IRB) or Medical Executive Committee, etc., and taking an individual's risk/benefit ratio into account.

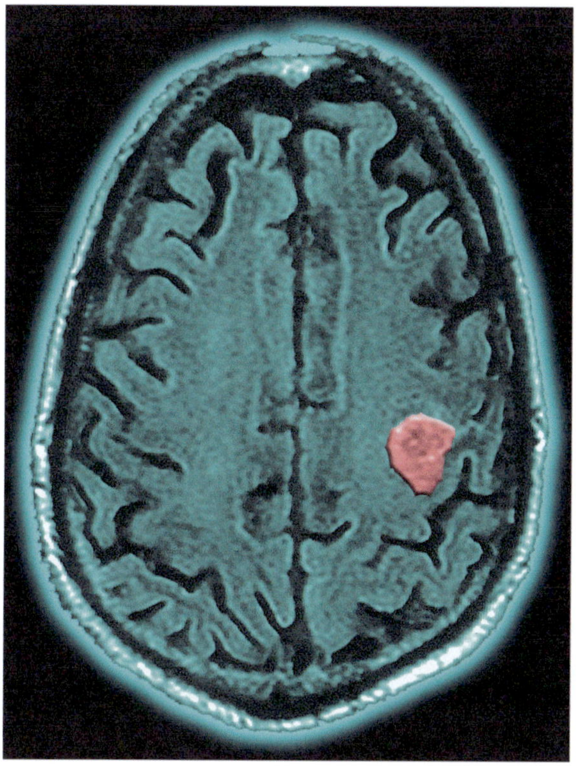

5. Drug/Medical Interactions

Several drug classes may lower seizure threshold. Therefore, it is recommended to conduct a complete medical history (see below) and determine an individual's risk/benefit ratio prior to administering TMS. Furthermore, certain drugs may interact with neuronal excitability and neurotransmitter function, which can alter the expected response to TMS and/or contaminate neurophysiologic measures (e.g., motor threshold or rTMS aftereffects).

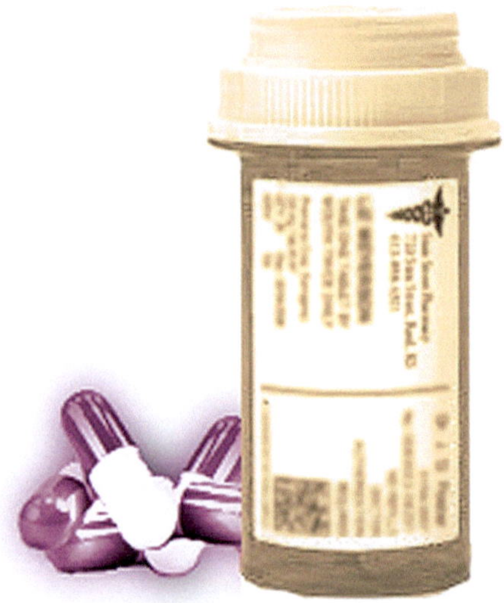

6. Recent Drug Withdrawal

Withdrawal from alcohol, barbiturates, benzodiazepines, or other anticonvulsants or sedatives may reduce seizure threshold. It is recommended that a complete medical history be determined.

7. Pregnancy

Fetal injury resultant from exposure to a TMS-generated magnetic field is highly unlikely. However, an induced seizure may adversely affect the fetus. It is recommended to take a conservative approach that takes the risk/benefit ratio into consideration. This may include administering a pregnancy test to individuals who are able to become pregnant prior to participation in a study.

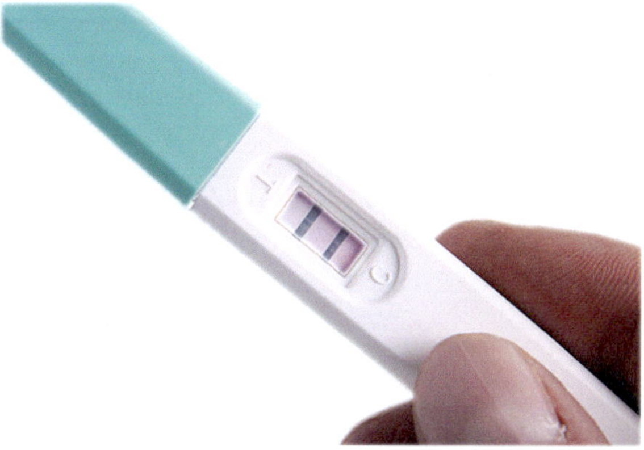

8. Children

No serious side effects of TMS have been reported to date in pediatric populations. Nonetheless, a conservative approach is recommended, and practitioners are encouraged to stay up to date on the pediatric TMS literature.

9. Illness

Since the effects of TMS may be influenced by baseline neural activity, it is important to be aware of any physical or mental illness which may alter this baseline activity.

2.2 Repetitive TMS Safety Guidelines

With regard to safe TMS delivery, there are four important parameters to consider: intensity (as % of resting motor threshold; RMT), frequency (in Hz), train duration, and inter-train interval. Although exceptions can be made in medically compelling instances, it is always best to approach each rTMS protocol conservatively by adhering to all published guidelines and approved protocols. The following tables list the recommended stimulation limits published by the US Food and Drug Administration (FDA) and common rTMS parameters endorsed by the Safety of TMS Consensus Group of the International Federation of Clinical Neurophysiology (IFCN).

Freq (Hz)	90% RMT	100% RMT	110% RMT	120% RMT	130% RMT	140% RMT	150% RMT	160% RMT	170% RMT	180% RMT	190% RMT	200% RMT
1	>1800	>1800	>1800	360	>50	>50	>50	>50	27	11	11	8
5	>10	>10	>10	>10	>10	7.6	5.2	3.6	2.6	2.4	1.6	1.4
10	>5	>5	>5	4.2	2.9	1.3	0.8	0.9	0.8	0.5	0.6	0.4
20	2.05	2.05	1.6	1.0	0.55	0.36	0.25	0.25	0.15	0.2	0.25	0.2
25	1.28	1.28	0.84	0.4	0.24	0.2	0.24	0.2	0.12	0.08	0.12	0.12

Maximum recommended stimulation duration (in seconds) of a single stimulation train

rTMS frequency	Number of studies	Average train duration	Average inter-train interval	Average number of trials
4–9 Hz	>10	Variable	Variable	Variable
10 Hz	>50	5–6 pulse-trains for 400–500 ms	3.2 s	250
20–25 Hz	>20	10 pulse-trains for 400–500 ms	17.1 s	80

Commonly recommended stimulation parameters supported by precedent

rTMS—Therapeutics: The effects of applying rTMS in repeated sessions over multiple days can be cumulative. It is important that therapeutic protocols remain within recommended safety limits and parameters are assessed following each session to avoid detrimental outcomes. Serious adverse events may still occur even from stimulation within guidelines; in this case, these events should be reported, and the protocol may need to be halted until a review can be performed.

rTMS—Cognitive research: Stimulation parameters should be established and approved prior to the commencement of research. It is important to understand that rTMS will not work the same in all participants and the protocol should not be amended to evoke a desired effect in "non-responding" participants. It is often recommended to "pilot" a new protocol to test the efficacy of the proposed parameters and determine if and how they influence the behavior being studied. When designing studies with multiple rTMS applications, it is important to be aware of the potential for carryover effects from prior sessions, which could last anywhere from 1 to 7 days.

rTMS—Motor cortex: It is recommended that rTMS to the primary motor cortex never exceed 130% RMT. If a participant's RMT cannot be determined, it is recommended that an intensity corresponding to the lower 95% confidence interval of the average RMT of other participants be utilized.

TBS: Given the ultra-high frequencies used (30–50 Hz) in TBS, extra care should be taken when choosing the appropriate dosing parameters (e.g., intensity and duration) to reduce the risk of an evoked seizure. When the modern TBS protocols were first described, they were applied using 80% of active motor threshold (AMT). This choice was reported to be based on an abundance of caution given limited safety data at the time on rTMS at frequencies over 20 Hz. However, there is evidence that AMT may reflect activity-dependent changes in excitability of spinal motor neurons more so than changes in cortical motor neurons. As such, RMT, rather than AMT, should be used to set the intensity of rTMS unless there is a clear reason to use AMT (e.g., RMT cannot be ascertained).

2.3 Seizure Recognition and Management

Induction of seizure, despite its rare occurrence (as of a 2021 Report by the Safety of TMS Consensus Group of the IFCN, a total of 41 seizures have been reported from rTMS), remains the most dramatic and medically dangerous acute complication of TMS. TMS pulses delivered at a high frequency with short inter-pulse and/or inter-train intervals are more likely to induce seizures. Although exceedingly rare, the risk of seizure must be addressed when discussing TMS with patients, participants, and ethics committee/IRBs, and a safety plan must be established prior to undertaking research or treatment. While most reported seizures occurred with rTMS, there have been isolated cases of seizures following single-pulse TMS in patient populations with heightened risk of seizure.

Seizures can manifest in different ways depending on the type of seizure and the circumstances, including the characteristics of the individual. Nonetheless, there are several common symptoms that are used to identify and recognize a seizure. These include:

- Staring blankly into space or sudden loss of consciousness
- Stereotyped motor activity, including chewing, mumbling, or picking at one's clothes
- Sudden incoherent actions, such as walking around, babbling, or removing clothing
- Sudden muscle stiffening, or rhythmic muscle jerks (generalized)

Medical attention should be sought for any suspected seizures. Furthermore, emergency medical attention should be sought if:

- The seizure lasts for at least 5 min.
- The individual is pregnant.
- The individual sustains injury.
- The individual suffers two seizures in quick succession.
- The individual does not regain consciousness.
- There is evidence of impaired cardiovascular activity (i.e., blue lips, lack of breathing, weak pulse).

Below, we will outline a basic treatment action plan. Your ethics committee/IRB may require a medical crash cart be present or immediately available in case of seizure.

2.3.1 *Steps for Responding to a Seizure and Administering First Aid*

Step 1: Stay calm and call for help. Document the exact time of seizure onset.

Step 2: If possible, recline the exam chair or help guide subject to the floor, roll the subject onto his/her side, and cushion the head.

Step 3: Loosen any restricting clothing around the airway, such as a tie

Step 4: After the seizure ends, document the exact time of seizure termination. Offer assistance and reassurance to the patient. Note that the patient may be disoriented and may not be aware that they had a seizure.

Under **NO** circumstances should you hold the subject down or put anything in their mouth.

Further Reading

Section 2.1

Horvath JC, Perez JM, Forrow L, Fregni F, Pascual-Leone A (2011). Transcranial magnetic stimulation: a historical evaluation and future prognosis of therapeutically relevant ethical concerns. *J Med Ethics*, 37(3), 137-43.

Pascual-Leone A, Houser CM, Reese K, Shotland LI, Grafman J, Sato S, Valls-Sol J, Brasil-Neto JP, Wassermannn EM, Cohen LG (1993). Safety of rapid-rate transcranial magnetic stimulation in normal volunteers. *Electroencephalogr Clin Neurophysiol*, 89, 120–30.

Rossi S, Hallett M, Rossini PM, & Pascual-Leone A (2009). Safety, ethical considerations, and application guidelines for the use of transcranial magnetic stimulation in clinical practice and research. *Clinical Neurophysiology*, 120(12), 2008-2039.

Wassermann EM, Grafman J, Berry C, Hollnagel C, Wild K, Clark K, et al. (1996). Use and safety of a new repetitive transcranial magnetic stimulator. *Electroencephalogr Clin Neurophysiol*, 101, 412–7.

Section 2.2

Di Lazzaro V, Restuccia D, Oliviero A, Profice P, Ferrara L, Insola A, Mazzone P, Tonali P, Rothwell JC. Effects of voluntary contraction on descending volleys evoked by transcranial stimulation in conscious humans. J Physiol (Lond). 1998 Apr 15;508 (Pt 2):625–33.

Fried PJ, Jannati A, Davila-Pérez P, Pascual-Leone A. Reproducibility of Single-Pulse, Paired-Pulse, and Intermittent Theta-Burst TMS Measures in Healthy Aging, Type-2 Diabetes, and Alzheimer's Disease. Front Aging Neurosci. 2017;9:263.

Guidance for Industry and Food and Drug Administration Staff Class II Special Controls Guidance Document: Repetitive Transcranial Magnetic Stimulation (rTMS) Systems. Issued on July 26, 2011. Available at: https://www.fda.gov/RegulatoryInformation/Guidances/ucm265269.htm

Horvath, J.C., Najib, U. (2014). Transcranial Magnetic Stimulation (TMS) Safety Considerations and Recommendations. In Pascual-Leone, A. (Ed.) *Transcranial Magnetic Stimulation, Neuromethods* (pp. 15-30). New York, NY: Springer Press.

Rossi S, Antal A, Bestmann S, Bikson M, Brewer C, Brockmöller J, Carpenter LL, Cincotta M, Chen R, Daskalakis JD, Di Lazzaro V, Fox MD, George MS, Gilbert D, Kimiskidis VK, Koch G, Ilmoniemi RJ, Pascal Lefaucheur J, Leocani L, Lisanby SH, Miniussi C, Padberg F, Pascual-Leone A, Paulus W, Peterchev AV, Quartarone A, Rotenberg A, Rothwell J, Rossini PM, Santarnecchi E, Shafi MM, Siebner HR, Ugawa Y, Wassermann EM, Zangen A, Ziemann U, Hallett M. Safety and recommendations for TMS use in healthy subjects and patient populations, with updates on training, ethical and regulatory issues: Expert Guidelines. Clinical Neurophysiology [Internet]. 2020 Oct 24 [cited 2020 Nov 30]; Available from: http://www.sciencedirect.com/science/article/pii/S1388245720305149

Rossi S, Hallett M, Rossini PM, & Pascual-Leone A (2009). Safety, ethical considerations, and application guidelines for the use of transcranial magnetic stimulation in clinical practice and research. *Clinical Neurophysiology*, 120(12), 2008-2039.

Wassermann EM (1998). Risk and safety of repetitive transcranial magnetic stimulation: report and suggested guidelines from the International Workshop on the Safety of Repetitive Transcranial Magnetic Stimulation, June 5–7, 1996. *Electroencephalogr Clin Neurophysiol*, 108, 1–16.

Section 2.3

Dobek, C. E., Blumberger, D. M., Downar, J., Daskalakis, Z. J., & Vila-Rodriguez, F. (2015). Risk of seizures in transcranial magnetic stimulation: a clinical review to inform consent process focused on bupropion. *Neuropsychiatric disease and treatment*, 11, 2975-87.

Rossi S, Antal A, Bestmann S, Bikson M, Brewer C, Brockmöller J, Carpenter LL, Cincotta M, Chen R, Daskalakis JD, Di Lazzaro V, Fox MD, George MS, Gilbert D, Kimiskidis VK, Koch G, Ilmoniemi RJ, Pascal Lefaucheur J, Leocani L, Lisanby SH, Miniussi C, Padberg F, Pascual-Leone A, Paulus W, Peterchev AV, Quartarone A, Rotenberg A, Rothwell J, Rossini PM, Santarnecchi E, Shafi MM, Siebner HR, Ugawa Y, Wassermann EM, Zangen A, Ziemann U, Hallett M. Safety and recommendations for TMS use in healthy subjects and patient populations, with updates on training, ethical and regulatory issues: Expert Guidelines. Clinical Neurophysiology [Internet]. 2020 Oct 24 [cited 2020 Nov 30]; Available from: http://www.sciencedirect.com/science/article/pii/S1388245720305149

Rossi, S., Hallett, M., Rossini, P. M., & Pascual-Leone, A. (2009). Safety, ethical considerations, and application guidelines for the use of transcranial magnetic stimulation in clinical practice and research. *Clinical Neurophysiology*, 120(12), 2008-2039.

Chapter 3
Setup and Basic Procedures

Abstract Laboratory and clinical settings commonly require the assessment of corticospinal excitability to the hand. The excitability can be quantified using the motor-evoked potential (MEP), which can be recorded using electrodes placed over an intrinsic hand muscle, and typically occur around 20 ms after a single suprathreshold pulse of TMS over the contralateral motor cortex. The technique requires correct placement of the electrodes over the peripheral muscle, and the TMS coil over the scalp, above the optimal stimulation site (hotspot). This section illustrates the procedure for electrode positioning and hotspot finding and describes assessment of the motor threshold, defined as the minimum stimulator intensity for a given coil type and device that can result in an MEP. This can be assessed under resting and active conditions. In this manual, we focus on the intrinsic hand muscles; however, these procedures can be adapted to many other muscles of the body. A visual perception known as *phosphenes* can occur with stimulation over the visual cortex and requires correct positioning and orientation of the coil. Phosphene hotspot identification and threshold method are described.

D. J. Edwards et al., *A Practical Manual for Transcranial Magnetic Stimulation*, https://doi.org/10.1007/978-3-031-62304-2_3

3.1 Attaching Electromyography Electrodes

A common peripheral measurement of the effect of TMS is the motor-evoked potential (MEP) (*Sect. 3.3*). MEP characteristics (such as amplitude and latency) can be measured using electromyography (EMG). For surface EMG, electrical potentials can be recorded from a muscle of interest with a negative, a positive, and a ground electrode. Below, we will demonstrate how to attach these electrodes to the first dorsal interosseous (FDI) muscle in a classic belly-tendon montage to measure the evoked activity from a TMS pulse to the contralateral motor cortex.

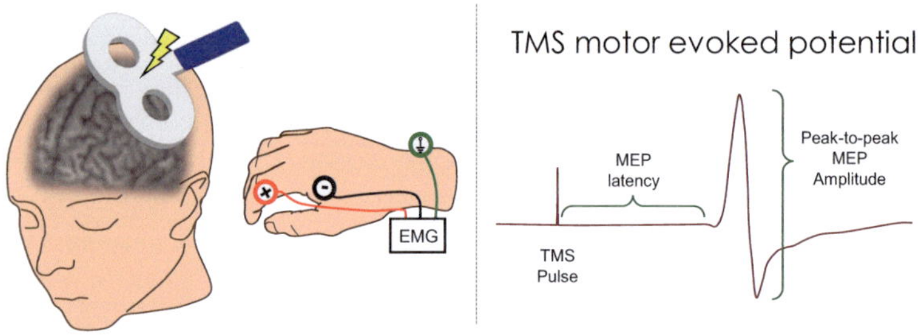

3.1.1 *Steps for Attaching EMG Electrodes*

Step 1: Locate the FDI muscle. To identify the belly of the FDI, ask subject to push index finger laterally against resistance: this will activate the FDI and make it easier to locate.

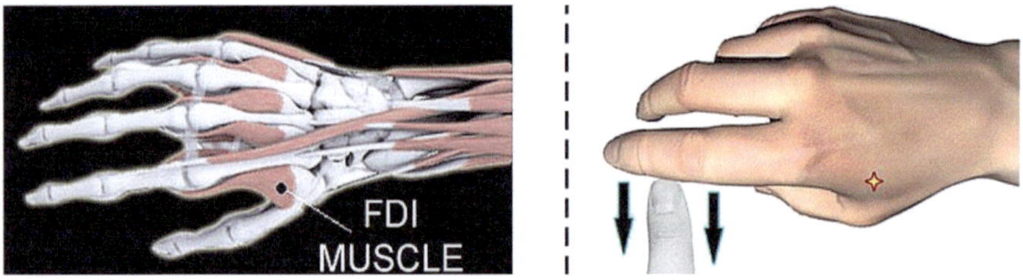

Step 2: Clean and lightly abrade skin over the FDI using alcohol or an alcohol wipe. The skin should be slightly pink afterward.

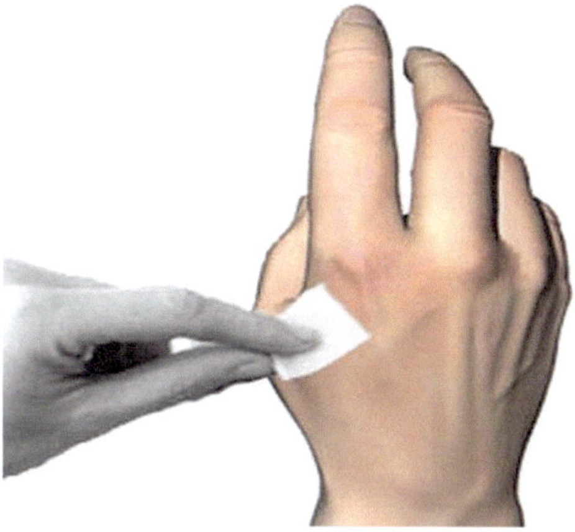

Step 3: Place a pea-sized amount of conductive gel on the face of the negative electrode. Place electrode face firmly against FDI muscle body and secure with medical tape.

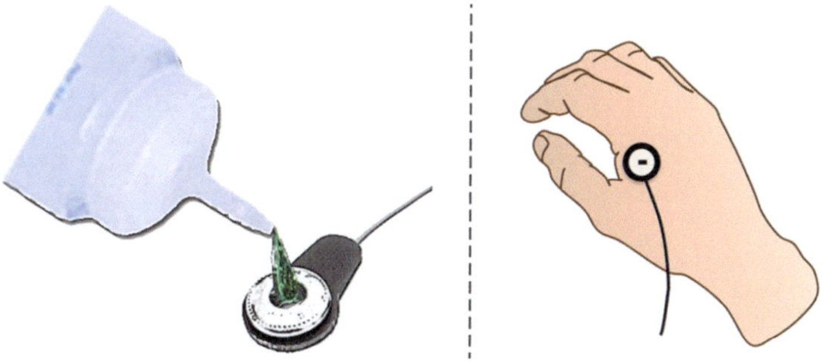

Step 4: Locate the first phalanx bone of the index finger. The tendon of the FDI muscle runs along this bone and attaches to the distal joint of the first phalanx.

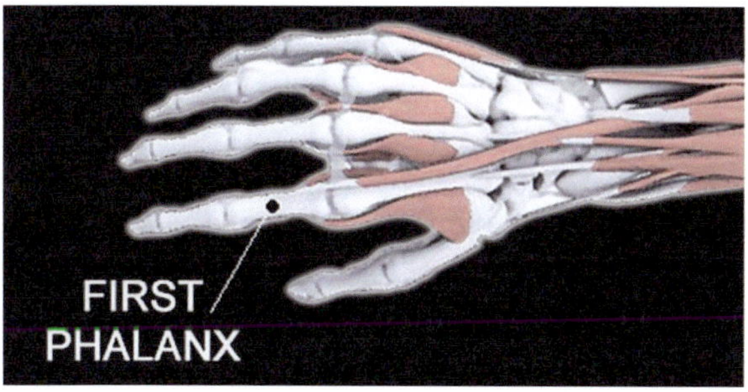

Step 5: Clean and lightly abrade skin over the lateral face of the first phalanx near the joint where the tendon attaches using alcohol or an alcohol wipe.

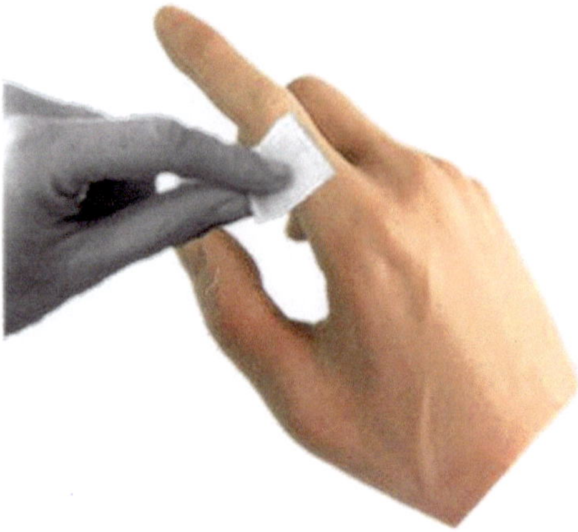

Step 6: Place a pea-sized amount of conductive gel on the face of the positive electrode. Place electrode face firmly against lateral face of the first phalanx distal joint and secure with medical tape.

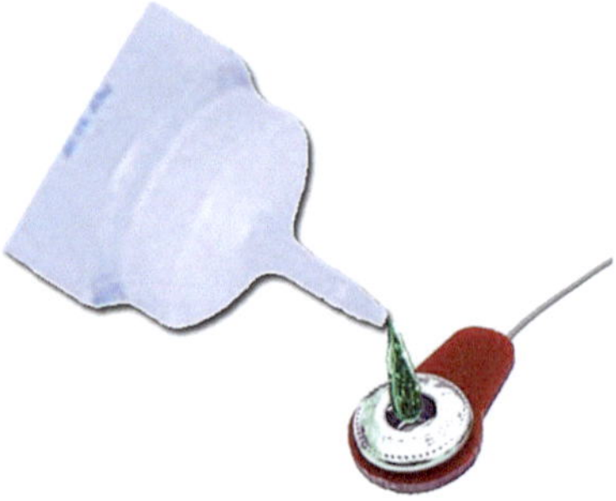

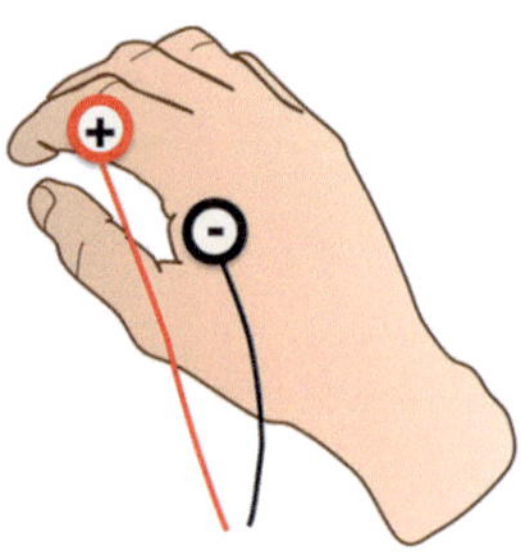

Step 7: Locate the ulnar head (also known as the ulnar styloid process). The ulnar head can be identified as a prominent bony bump on the dorsomedial wrist below the pinky.

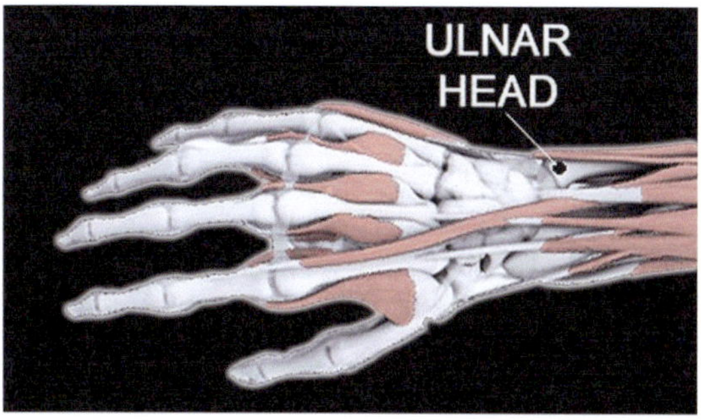

Step 8: Clean and lightly abrade skin over the ulnar head using alcohol or an alcohol wipe.

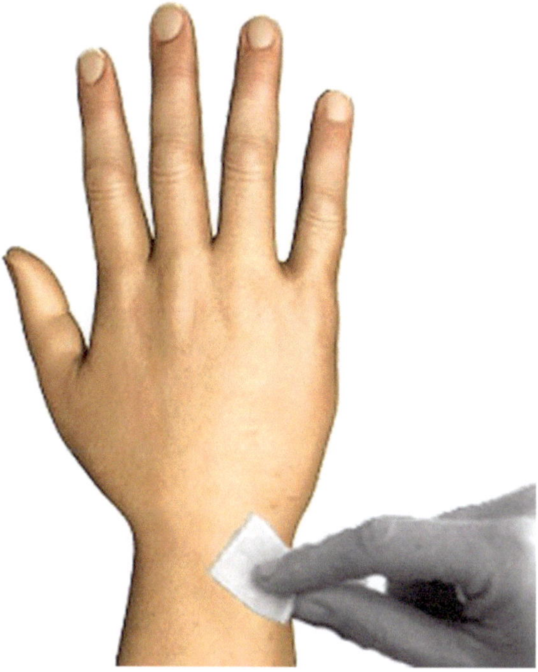

Step 9: Place a pea-sized amount of conductive gel on the face of the ground electrode. Place electrode firmly against the ulnar head and secure with medical tape.

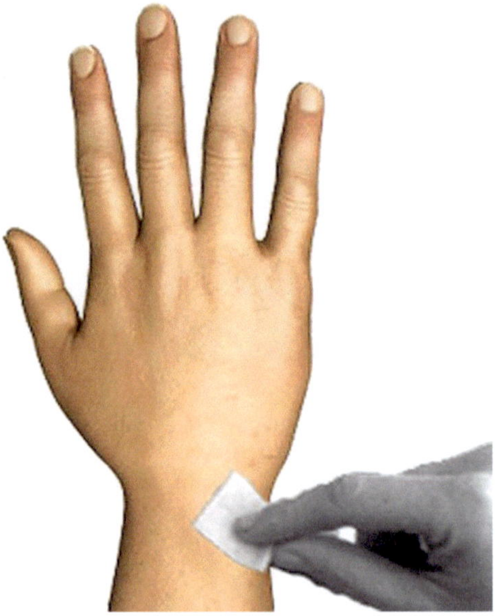

Notes:

- EMG recording is not restricted to the FDI. As shown below, other commonly used muscles include the abductor pollicis brevis (APB; with the positive electrode placed on the lateral face of the first phalanx of the thumb) and the abductor digiti minimi (ADM; with the positive electrode placed on the medial face of the first phalanx of the 5th digit/little finger). Additionally, the ground electrode can be placed on the flexor carpi radialis or any meaty portion of the forearm.

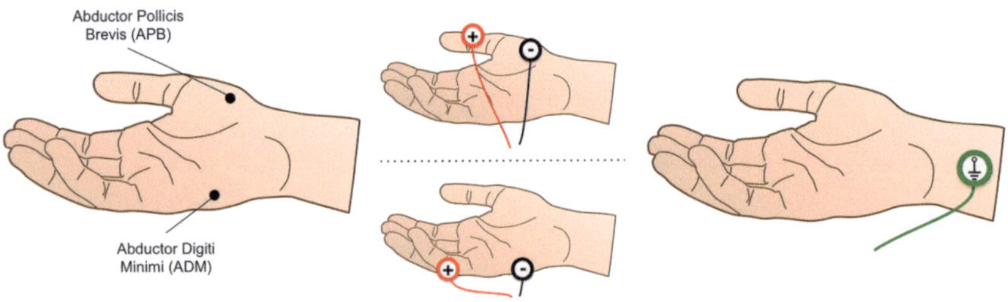

- Although we show surface electrodes in the above example, needle electrodes can also be used for the same purpose.
- It is good practice to standardize electrode placement across participants and visits on the same participant.

3.2 Finding the Cranial Vertex

This section covers how to locate the cranial vertex (also referred to as Cz in the International 10-20 System for EEG electrode placement). The vertex serves as a starting point for many exploratory TMS procedures, including determining the motor hotspot (*Sect. 3.3*).

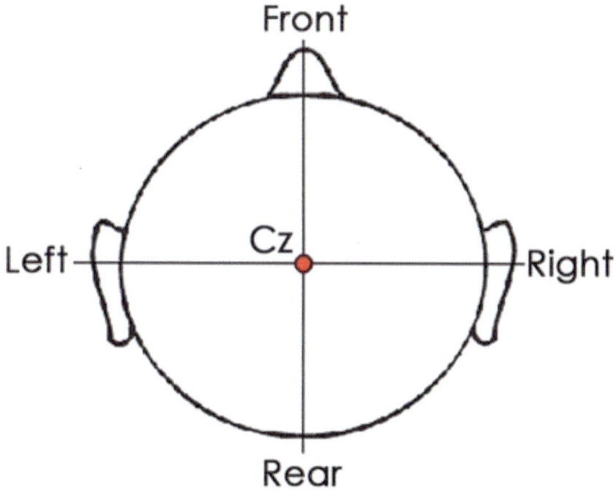

3.2.1 Steps for Finding the Cranial Vertex

Step 1: Locate the nasion (the indentation above the bridge of the nose) and the inion (the ridge on the occipital bone). If the inion is not prominent, look for the change in angle of the occipital bone from the neck.

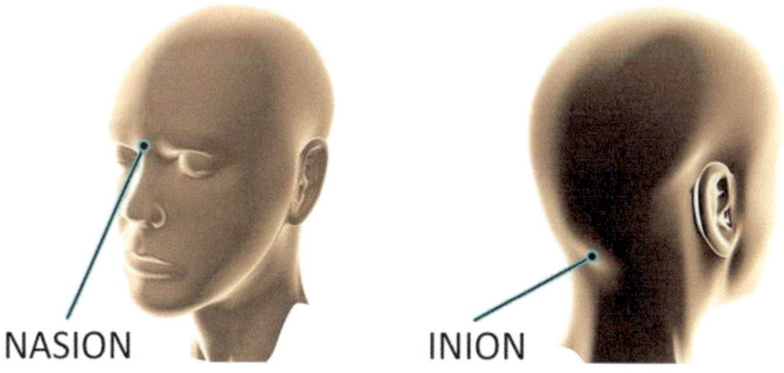

Step 2: Measure the distance between the inion and nasion along the mid-sagittal line over the cranium. Mark the midpoint of this line; this is the horizontal axis of the vertex.

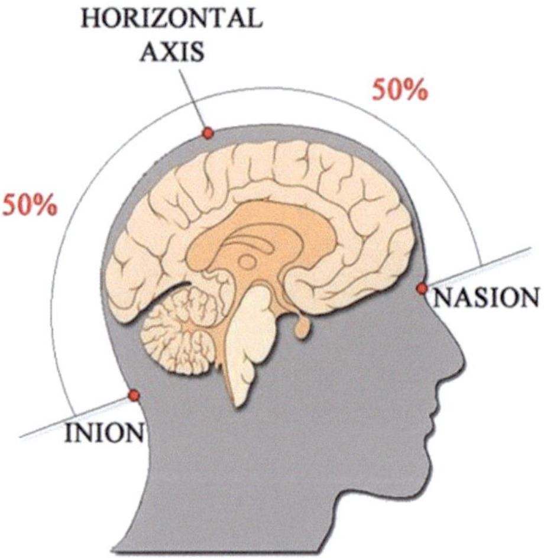

Step 3: Locate the left and right tragus (the cartilage immediately in front of the ear canal). The same part of the left and right tragus should be used.

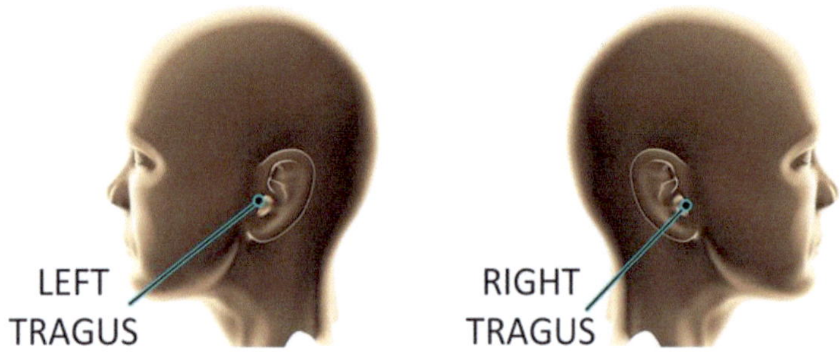

Step 4: Measure the distance between the left and right tragus over the cranium. Make sure the line crosses the point from Step 3. Mark the midpoint of this line; this is the vertical axis of the vertex.

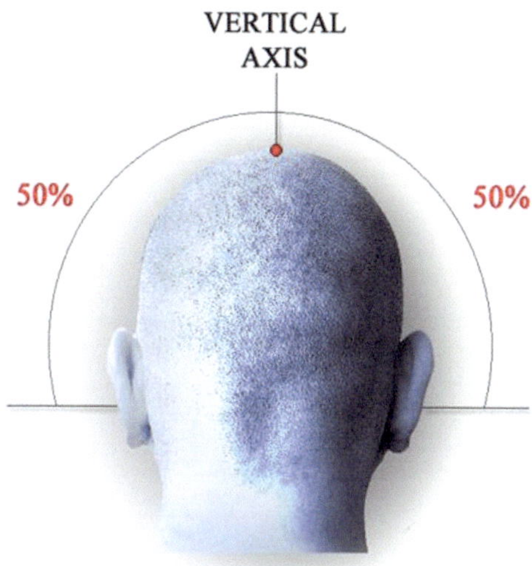

Step 5: The cranial vertex, or Cz, is the point of intersection between the vertical and horizontal axes.

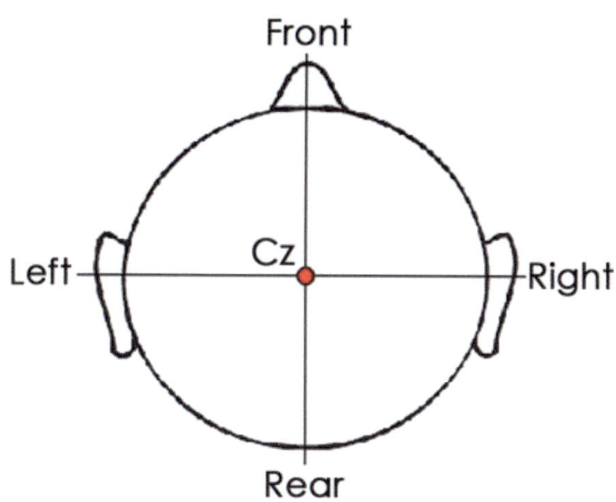

Notes:

- To make this process easier, all points can be marked onto a snug-fitted swimming cap or directly on the participant's scalp using a grease pencil or washable marker.
- In addition, the location of the eyebrows and the distance between the cap and the eyebrows can be measured and marked to check that the cap does not move during the session.

3.3 Locating the Motor Hotspot

The primary motor cortex (M1) contains a broad somatotopic representation of the different muscles of the body. Using TMS, the musculature representations can be selectively stimulated, and brief contractions of the targeted muscle can be elicited. The characteristics of the elicited muscle response—called MEPs—can be measured using EMG (*Sect. 3.1*) and serve as a quantification of the TMS effect.

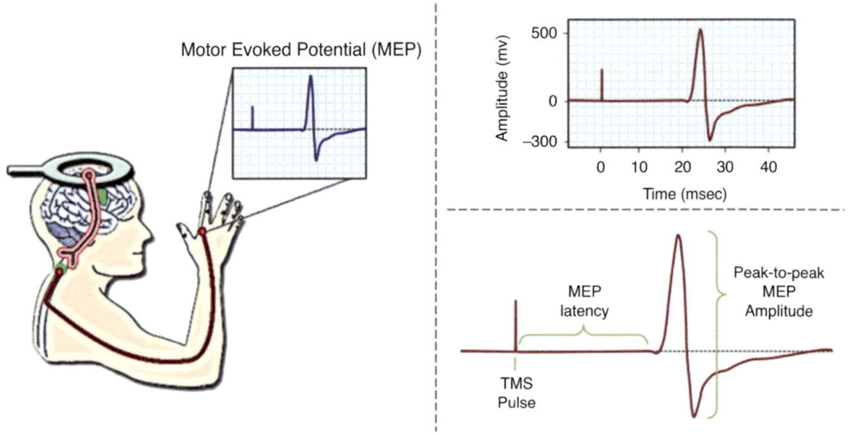

Due to the spread of the electromagnetic field (*Sect. 1.5*), the focality of TMS is typically larger than the representation of a single muscle (especially at higher intensities). Therefore, it is necessary to find the optimal coil position for a given muscle rather than any coil position that elicits an MEP or visible contraction. The optimal coil position, termed the motor "hotspot," is defined as the precise coil location and orientation that elicits the largest and most consistent response from the target muscle.

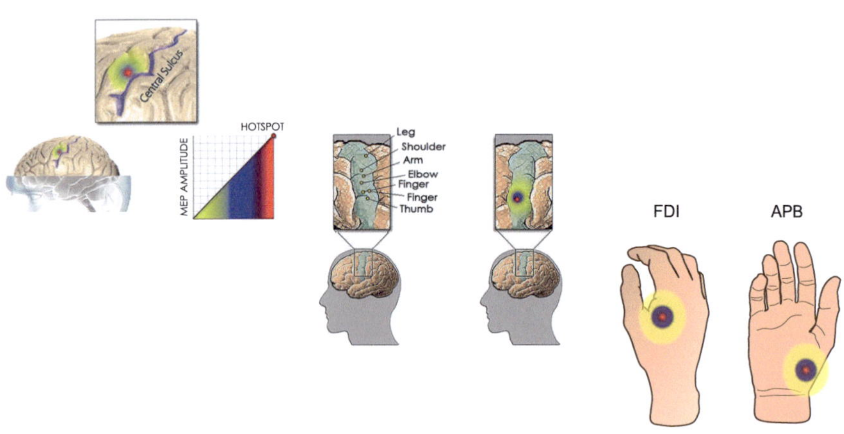

In this section, we will determine the hotspot for the FDI muscle of the right hand. As muscles are controlled by the contralateral hemisphere, this location will be in the left M1. The same approach can be applied to any other muscle, for example, APB.

3.3.1 *Steps for Locating the Left Motor "Hotspot" with the Right FDI as the Target Muscle*

Step 1: Attach EMG electrodes to the right FDI using a belly-tendon montage with a ground on the bony wrist (*Sect. 3.1*).

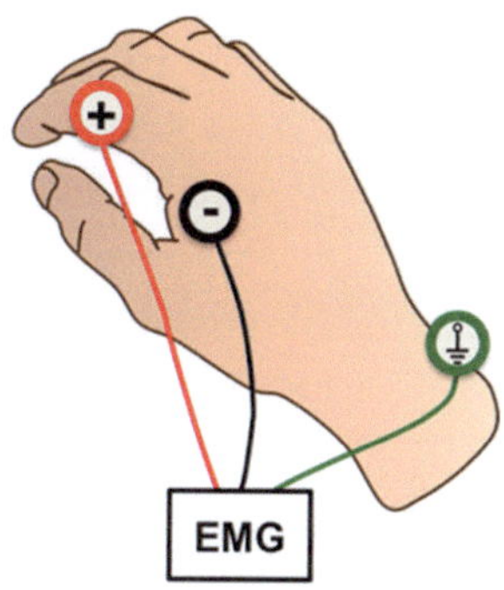

Step 2: Find the cranial vertex/Cz and mark it on a swim cap (*Sect. 3.2*). In this example, this location is marked "A."

Step 3: From the vertex, measure and mark points 5 cm dorsal and 5 cm lateral. In this example, these points are marked, "B" and "C," respectively. Point C is the starting point for the search of the motor hotspot; this is where you will place the center of the coil.

Step 4: Draw a dotted line connecting points B and C. This line will help you orient the coil 45° from the midline (or roughly perpendicular to the central sulcus).

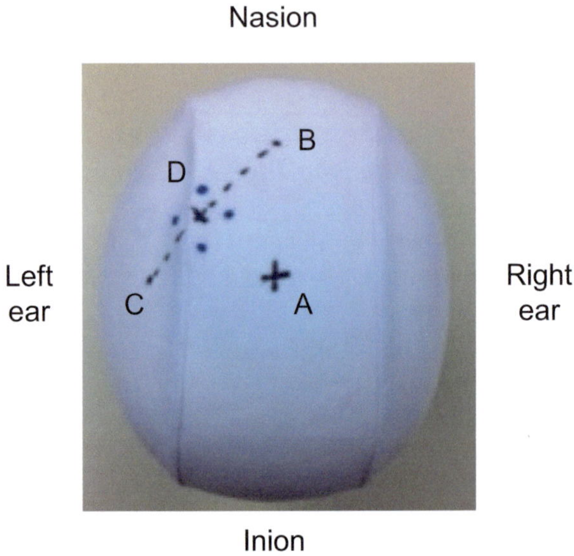

Step 5: To provide a visual marker that tells you where the center of the coil is, measure the distance between the center and the front of the TMS coil (or consult the coil specifications manual). In this example, the distance between the center and the front of the coil is 3.5 cm.

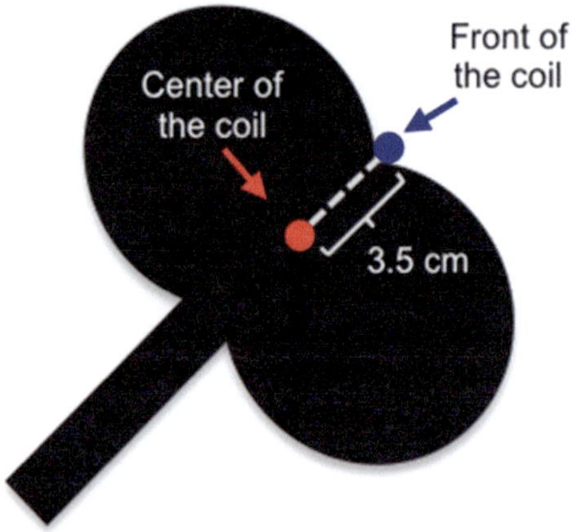

Step 6: Next, mark point "D" using the measurement from Step 5 (3.5 cm in this example). This represents where we will place the front of the coil. Measure this distance from point C to point B (along the line from *Step 4*).

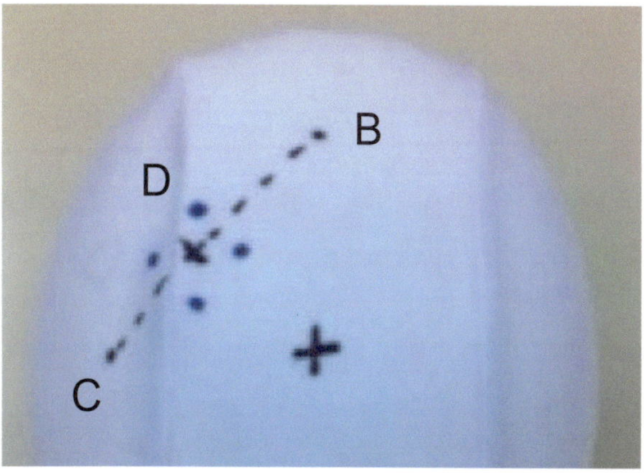

Step 7: Place the front of the TMS coil at point D with the handle angled along the dotted line from Step 4. This should put the center of the coil (magnetic peak) over point C, which is the starting location to find the left FDI hotspot, with a 45° angle from the mid-sagittal line. The coil should be held tangentially to the scalp.

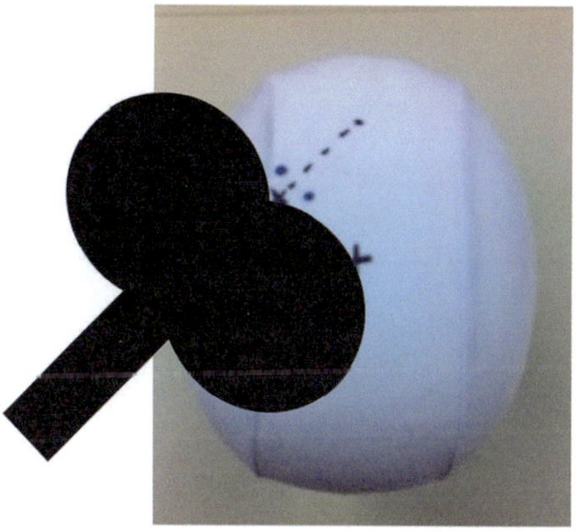

Step 8: Set initial TMS output to a very low intensity (lower than the intensity that elicits responses). In this example, we use 30% MSO. Maintain the same coil location and position as *Step 7* and deliver three pulses spaced approx. 6–10 s apart.

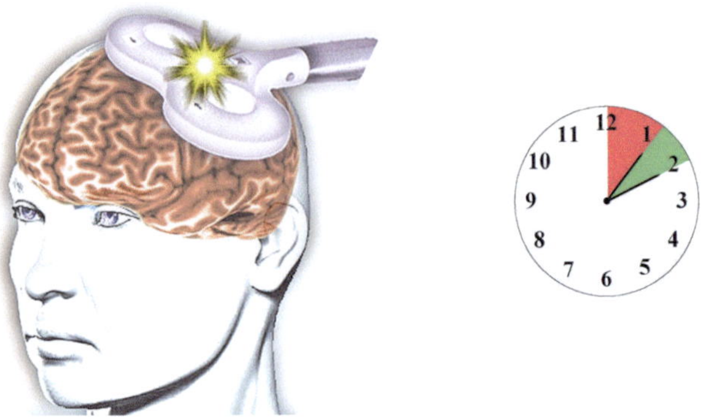

Step 9: If you a visible muscle twitch is not observed, increase intensity by 10 (e.g., 40% MSO) and repeat from *Step 8* until a visible twitch is observed in the target muscle.

Step 10: When a visible muscle twitch is observed, check the EMG system to obtain the peak-to-peak amplitude of the corresponding MEP. Note that the latency of the MEP (time between the TMS pulse artifact and the start of the MEP) should be approximately 20 ms).

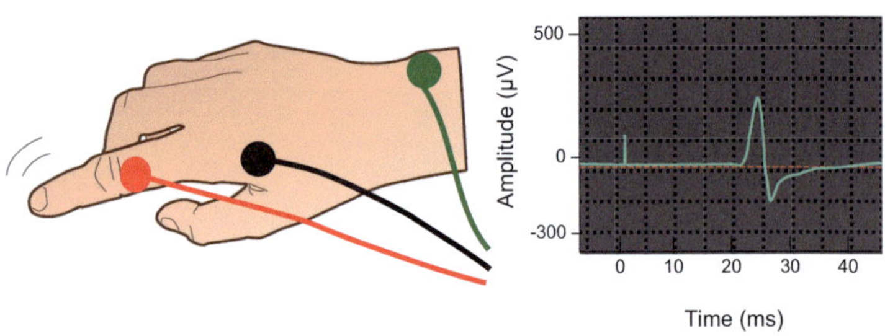

Step 11: Repeat each TMS pulse at least three times before changing intensity or location.

Step 12: Now that the intensity is high enough to elicit a muscle response from the starting location, we will keep intensity constant and compare the responses from different locations. From point D, measure 1 cm anterior, medial, posterior, and lateral and mark these points on the cap (shown here in blue).

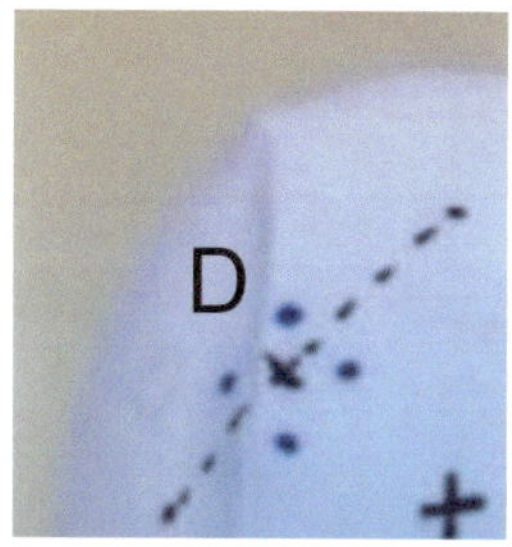

Step 13: Begin with the anterior spot. Move the interior curve of the TMS coil to this point and align the handle parallel to the 45° angle form *Step 4*.

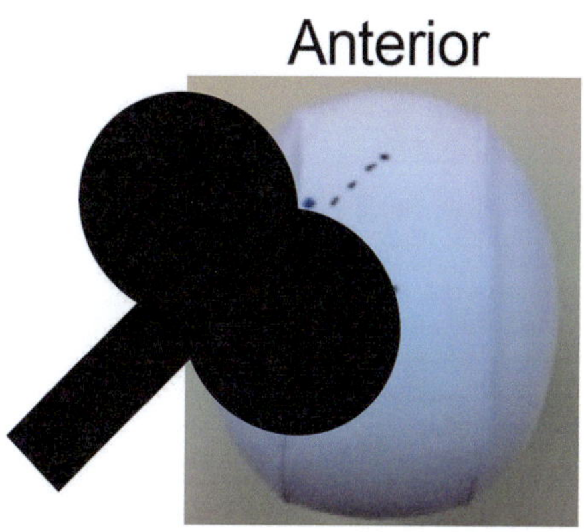

Step 14: Keeping intensity constant, give three single pulses (6–10 s apart) at this new location. For each pulse, first look for visible twitch of the target muscle and then check the MEP peak-to-peak amplitude.

Step 15: Repeat *Step 14* at each of the other three points, moving clockwise (e.g., medial, posterior, lateral).

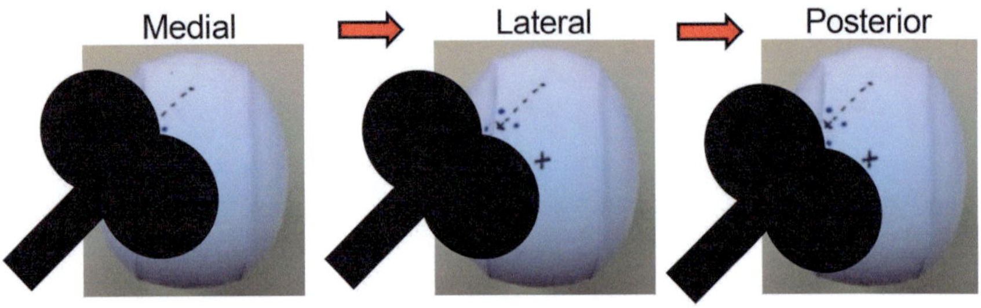

Step 16: Compare the MEPs recorded at each location and determine which of the five locations (center, anterior, medial, posterior, lateral) elicits the largest and most consistent MEPs along with visible twitch of the target muscle. **This position is the hotspot.**

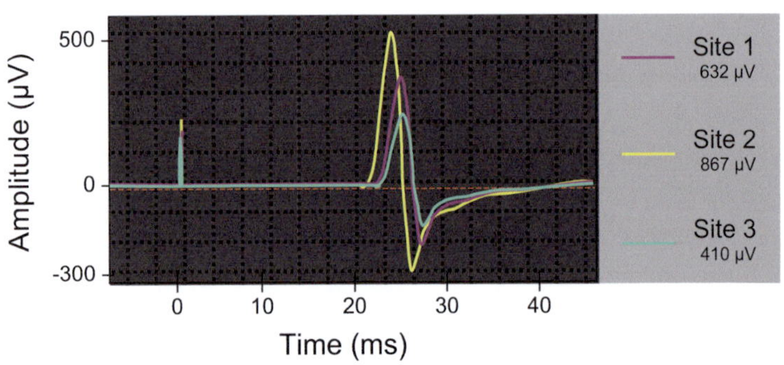

Notes:

- Remember, the hotspot is the location under the center of the coil. The front of the coil is just a visual reference to help in placing the coil.
- The number and location of additional search points (e.g., anterior, medial, posterior, lateral) is one example to find the motor hotspot quickly and easily. A more complex grid or additional search points may be used for a more precise assessment.
- Search locations including the final hotspot can either be marked on a snugly fitted swim cap or directly onto the scalp using either a grease pen or light washable marker.

3.4 Determining the Resting Motor Threshold

Motor threshold (MT) is the measurement of an individual's motor cortex excitability (or reactivity). MT is used in nearly every TMS setting (including clinic and basic research) to individualize the intensity of TMS across individuals and is specific to the TMS stimulator and coil. MT can be measured in a muscle that is relaxed or contracted. The latter is called the AMT and will be discussed in *Sect. 3.5*. The former is called the RMT and is the focus of this section.

RMT is defined as the *minimum* intensity of TMS necessary to elicit a motor response 50% of the time when the target muscle is at rest. The excitability assessed by RMT is believed to reflect, in part, the activity of voltage-gated ion channels in M1 neurons. Every motor response from a TMS pulse to M1 is an interaction between the induced current and the basal excitability of the neurons. This relationship is further influenced by both technical parameters and individual factors. To account for technical parameters (such as TMS device, coil design, etc.), it is important to always measure the RMT separately for each TMS machine and coil. Even with the same equipment, there is high inter-individual variability in RMT, due in part to stable (or non-modifiable) factors, such as age, sex, or scalp-to-cortex distance. Accordingly, there is a very wide range in RMT values across the population. Within a subject, RMT is typically stable over time and may be the most reliable TMS measurement. However, as the excitability of the brain oscillates, there can be trial-to-trial variation in MEP amplitude. For this reason, the IFCN recommends using a series of 10 pulses per intensity step when determining the RMT. This method is the most widely used approach.

IFCN recommendations define the RMT as *the minimum TMS intensity necessary to elicit a MEP ≥ 50 μV (peak-to-peak amplitude) in at least 5 of 10 pulses.*

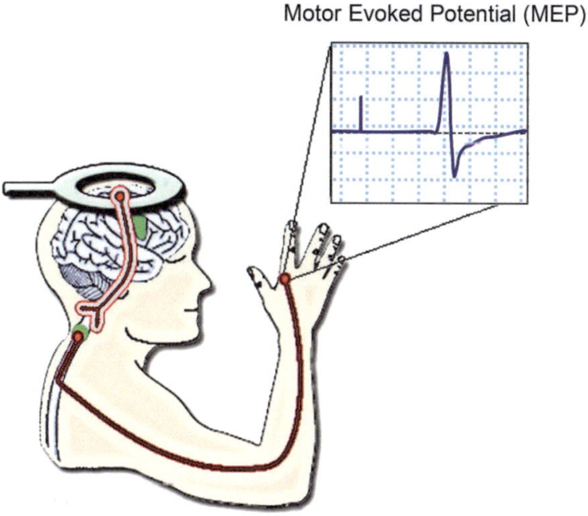

This section will detail the steps to assess the RMT for the right FDI (left M1).

3.4.1 Steps for Determining the Resting Motor Threshold for the Right FDI

Step 1: Attach EMG electrodes to the right FDI using a belly-tendon montage with a ground on the bony wrist (*Sect. 3.1*).

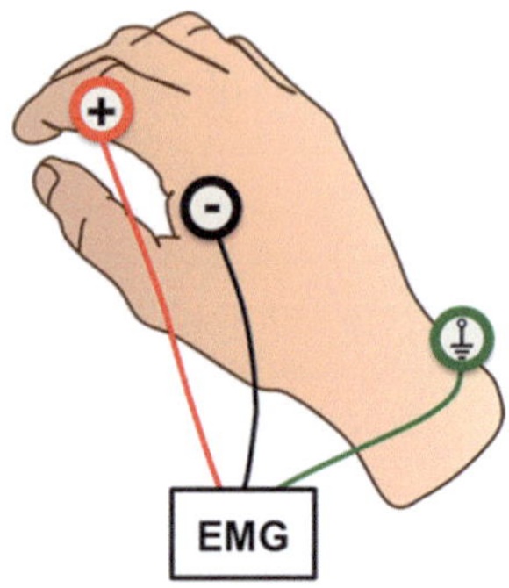

Step 2: Determine the FDI motor hotspot (*Sect. 3.3*). Remember that when determining the hotspot, one should start at a low intensity (i.e., below threshold) and increase until responses are large and consistently observed (i.e., above threshold). The hands should be completely at rest (i.e., no EMG activity) for the hotspot.

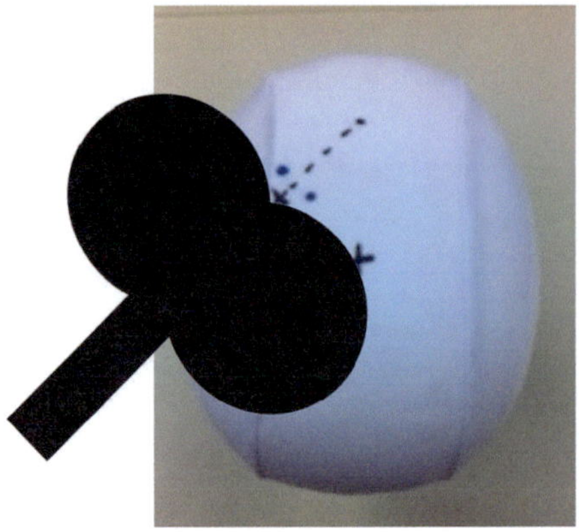

Step 3: Since the intensity reached by the end of Step 2 is high enough to consistently elicit responses (i.e., above threshold), we expect our RMT to be lower. In this example (see figures), we saw no MEPs at 30% MSO, small (<50 μV) and inconsistent MEPs at 40% MSO, and large (>50 μV) and consistent MEPs at 50% MSO.

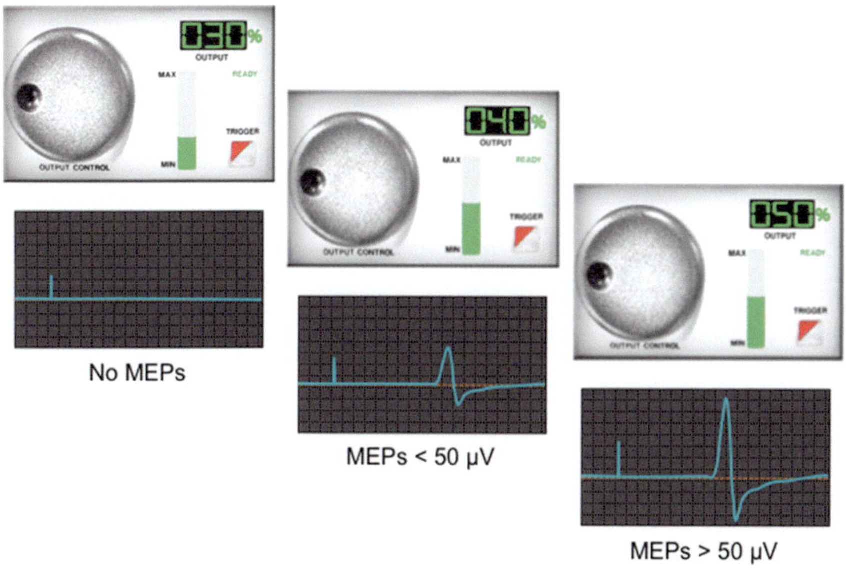

Step 4: Deliver 10 TMS pulses (spaced 6–10 s apart to avoid cumulative effects). For each trial, note the MEP amplitude. In this example, all MEPs at 50% MSO are ≥50 μV indicating that the stimulation intensity is above threshold.

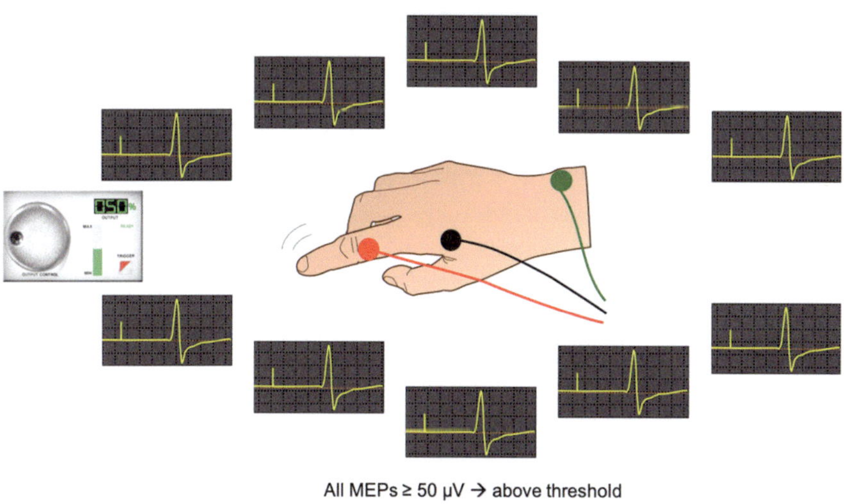

Step 5: If more than 5 of 10 TMS trials elicit MEPs ≥50 µV, reduce intensity by 2% MSO and repeat *Step 4*.

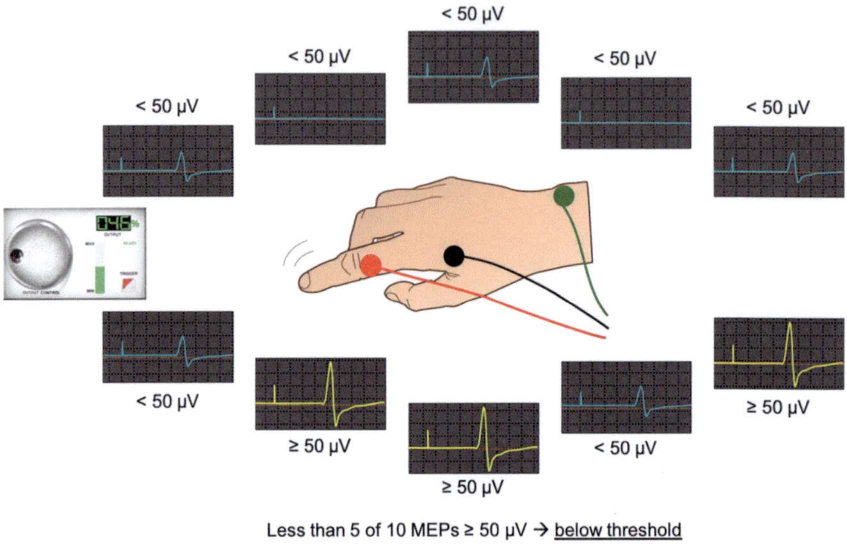

Step 6: Repeat Steps 4 and 5 until reaching an intensity where fewer than 5 of 10 TMS trials elicit an MEP ≥50 µV (in this example, we are at 46% MSO and observe MEPs ≥50 µV in 3 of 10 trials). This indicates we are below threshold.

Step 7: Increase intensity by 1% MSO and repeat Step 4. If at least 5 of 10 TMS trials elicit an MEP ≥50 μV you have reached the threshold; if not, the threshold is 1% MSO higher (i.e., the last intensity step where at least 5 of 10 MEPs were ≥50 μV). **In this example, the threshold is 47% MSO.**

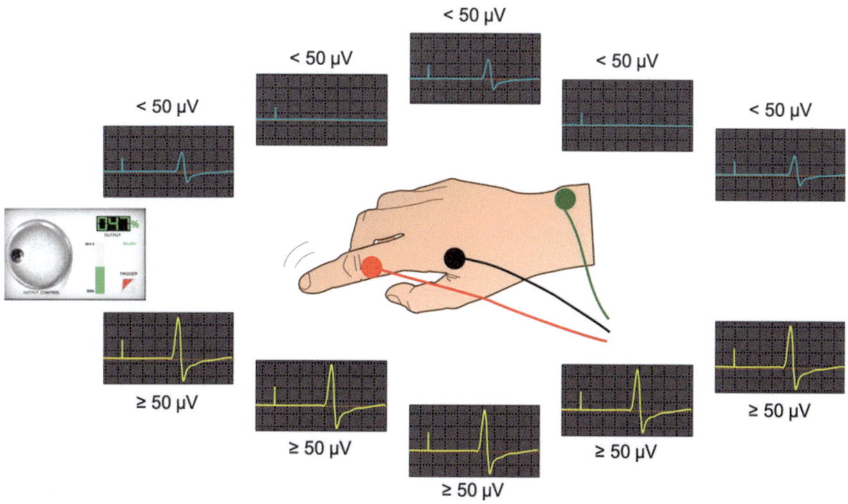

At least 5 of 10 MEPs ≥ 50 μV → at threshold

Notes:

- It is common for the first MEP of any block to be heightened due to anticipation or startle.
- If MEPs are very large (e.g., >1 mV), you can decrease intensity by more than 2% MSO in *Step 5*.
- When time constraints are an issue, this procedure can be shortened by using a criterion of three of six MEPs ≥50 μV in *Steps 4–7* above (instead of 5 of 10). It is important to note, though, this method has not been formally validated.
- It is important the subject keep his/her hand relaxed during this procedure. In addition, the subject should minimize all movement (including speech) and keep his/her eyes open.
- It is important to minimize movements of the TMS coil and the head during assessment of the RMT as even small movements can change the TMS trajectory and influence the measurement.
- If you do not have EMG measuring equipment, the above procedure can be conducted using ***relative visual inspection***. With each pulse, the TMS administrator should visually watch and note whether there was a resultant contraction in the target muscle. The lowest intensity that elicits at least 5 out of 10 visible muscle twitches is the resting motor threshold.

3.5 Determining the Active Motor Threshold

AMT is obtained in much the same manner as RMT (*Sect. 3.4*), the only difference being that the target muscle is isometrically contracted rather than relaxed. Muscle contraction pre-activates the entire descending motor system, somewhat enhancing the excitability of at multiple levels of the corticospinal pathway.

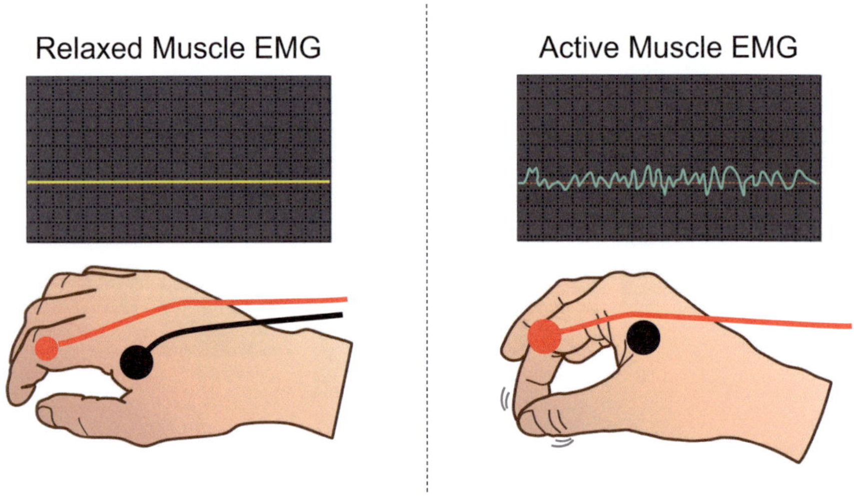

The AMT will always be lower than the RMT as the increase in excitability from muscular activation means *less* cortical stimulation (i.e., lower TMS intensity) is required to generate an MEP. As shown in the figure below, MEPs elicited during activation of the target muscle are usually followed by a period of reduced or absent voluntary muscle activity, which is termed the silent period (*Sect. 4.2*).

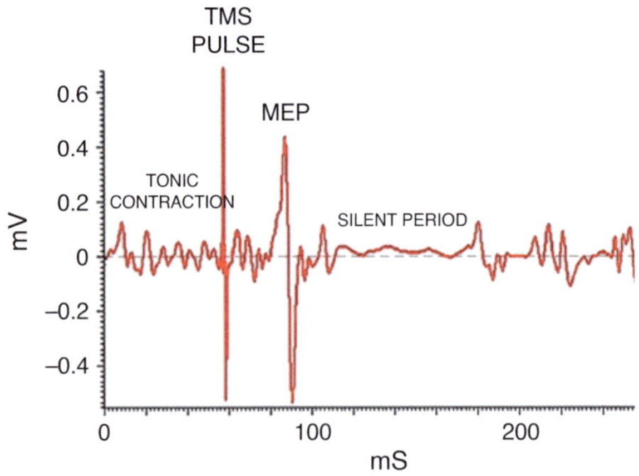

The majority of TMS protocols use the subject's RMT to set the intensity of subsequent stimulation. However, there are a few circumstances where AMT may be preferred:

- Where RMT cannot be reliably determined (i.e., >100% MSO) due to reduced cortical excitability.
  - Special populations (e.g., young children and in certain neurologic diseases or disorders).
  - Certain medications (e.g., ion channel-blocking anticonvulsants) or recreational drugs (e.g., acute alcohol intoxication).
- When examining the direct effects of TMS on specific motor movements.
  - Since TMS will be applied during ongoing muscle activity, the AMT may be a better index of the "active" state of the motor system than RMT.
- When there is an elevated risk of a serious adverse event, such as a seizure.
  - Historically, novel rTMS protocols (i.e., TBS; *Sect. 6.1*) have used AMT since it will lead to a lower stimulation intensity than RMT.
  - Certain medications or drugs (e.g., stimulants, alcohol withdrawal).

This section will detail the steps to determine AMT in the contracted right FDI (left M1) following IFCN recommendations.

3.5.1 *Steps for Determining the Active Motor Threshold for the Right FDI*

Step 1: Attach EMG electrodes to the right FDI using a belly-tendon montage with a ground on the bony wrist (*Sect. 3.1*)

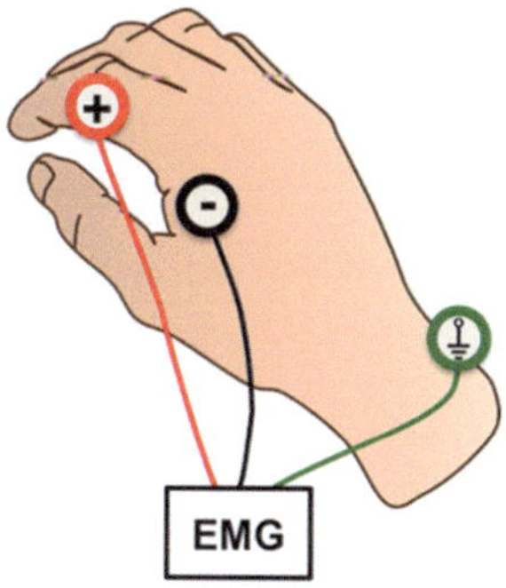

Step 2: Determine the FDI motor hotspot (*Sect. 3.3*). Remember that when determining the hotspot, one should start at a low intensity (i.e., below threshold) and increase until responses are large and consistently observed (i.e., above threshold).

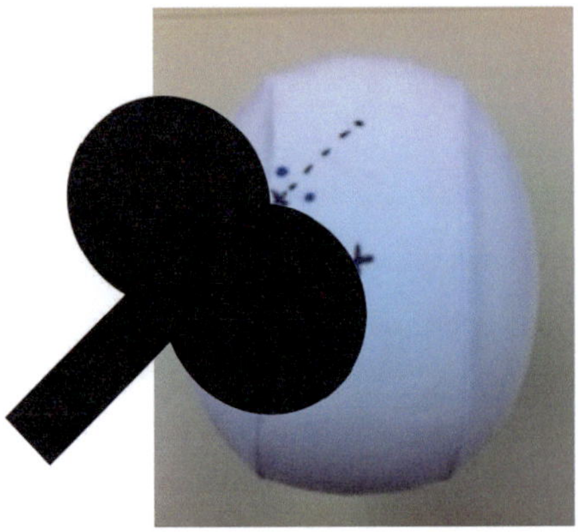

Step 3: Ask subject to gently contract the FDI muscle. This can be accomplished by holding the index finger against gravity or by gently squeezing it against the thumb in a light pinch grip. The target level of activation is about 20% of maximum contraction. If a force transducer is not available, ask the subject to maintain about 100–200 μV of muscle contraction, using the live EMG recording as feedback.

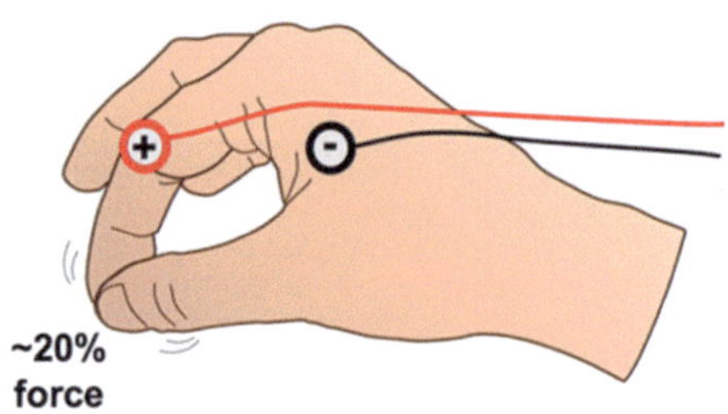

Step 4: If you previously determined RMT (*Sect. 3.4*) you can use this as the starting intensity, since AMT will always be lower, and skip to Step 6. Otherwise, set initial TMS output to a very low intensity. In this example, we use 30% MSO. Give 2–3 pulses (6–10 s apart) and record the resulting MEPs if present.

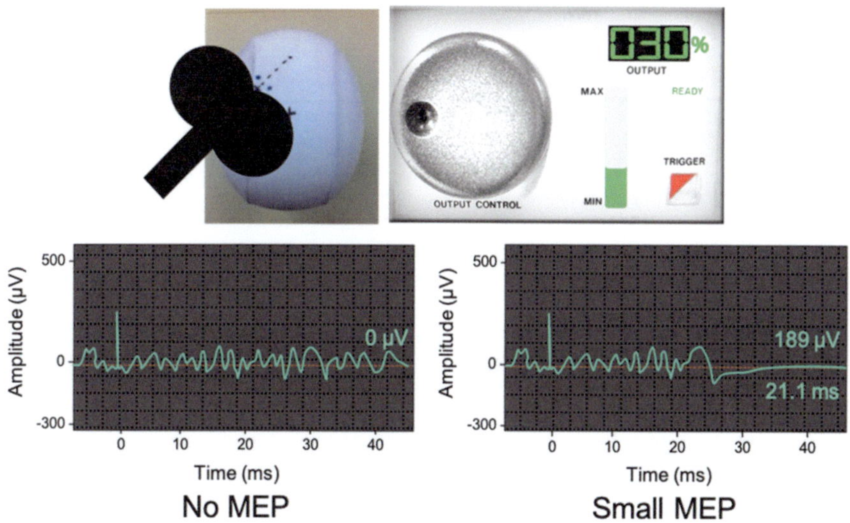

Step 5: If MEPs are absent or below 200 μV amplitude, increase the intensity progressively in steps of 10% MSO, until MEPs ≥200 μV are observed. Give 2–3 pulses at each intensity step and wait 6–10 s between stimuli (to avoid cumulative effects). In this example (see figures), we saw small (<200 μV) and inconsistent MEPs at 30% MSO and large (>200 μV) and consistent MEPs at 40% MSO.

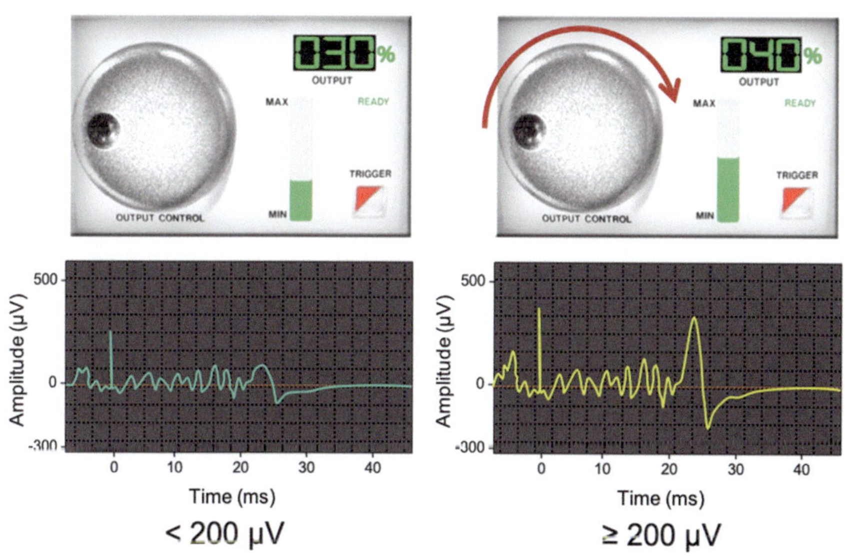

Step 6: Deliver 10 TMS pulses (spaced 6–10 s apart to avoid cumulative effects). For each trial, note the MEP amplitude. In this example, at 40% MSO, all MEPs are ≥200 μV. This indicates you are above threshold.

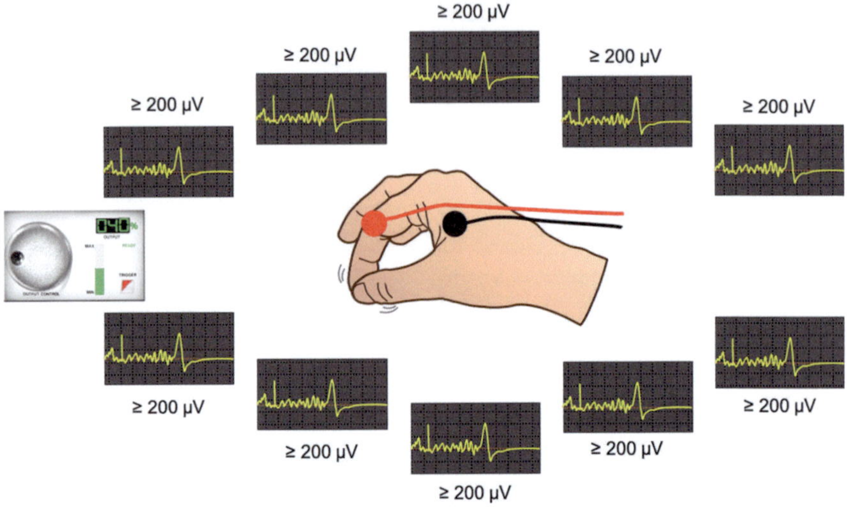

All MEPs ≥ 200 μV → above threshold

Step 7: If more than 5 of 10 TMS trials elicit MEPs ≥200 μV, reduce intensity by 2% MSO and repeat Step 6.

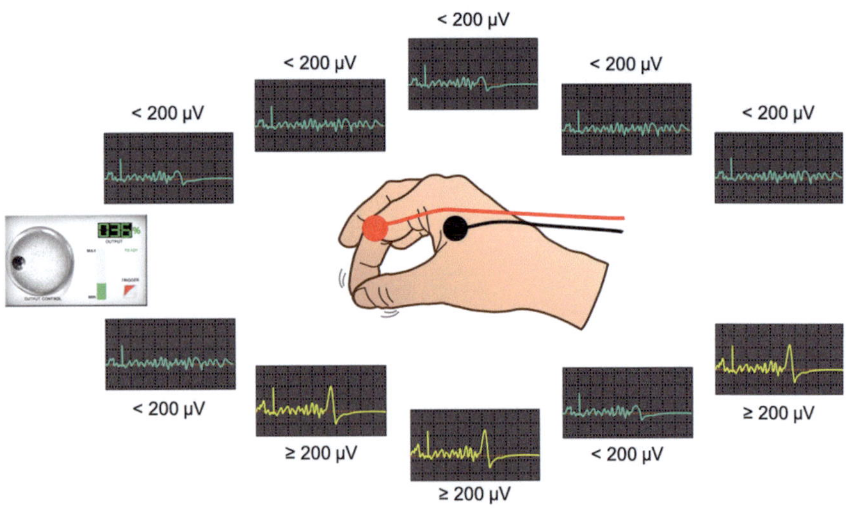

Less than 5 of 10 MEPs ≥ 200 μV → below threshold

Step 8: Repeat Steps 6 and 7 until reaching an intensity where fewer than 5 of 10 TMS trials elicit an MEP ≥200 μV (in this example, we are at 36% MSO and observe MEPs ≥200 μV in 3 of 10 trials). This indicates you are below threshold.

Step 9: Increase intensity by 1% MSO and repeat Step 6. If at least 5 of 10 TMS trials elicit an MEP ≥20 μV you have reached the threshold; if not, the threshold is 1% MSO higher (i.e., the last intensity step where at least 5 of 10 MEPs were ≥20 μV). In this example, the threshold is 37% MSO.

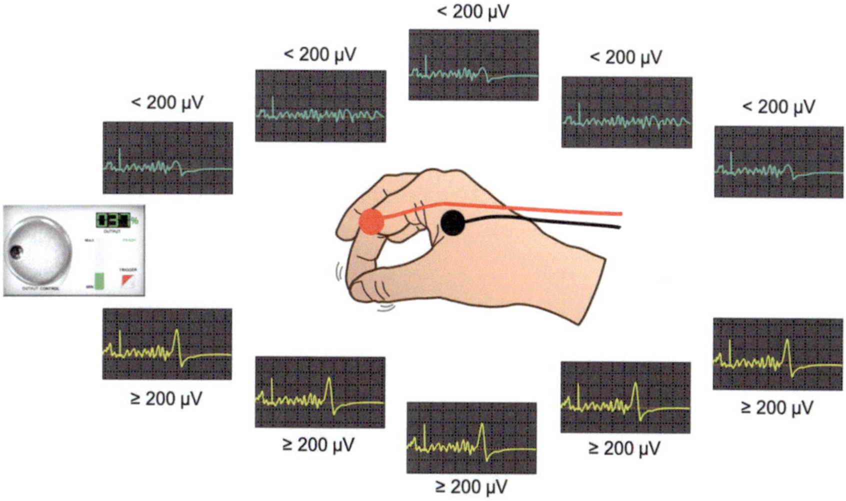

Notes:

- It is common for the first MEP of any block to be heightened due to anticipation or startle.
- If MEPs are very large (e.g., >1 mV), you can decrease intensity by more than 2% MSO in Step 5.
- When time constraints are an issue, this procedure can be shortened by using a criterion of three of six MEPs ≥200 μV in *Steps 6–9* above (instead of 5 of 10). It is important to note, though, this method has not been formally validated.
- It is important that the subject maintains a steady contraction of the target muscle during this procedure. A continuous EMG output that is visible to the participant can aid in this. It may help to allow the participant rest between blocks. In addition, the subject should minimize all other movement (including speech) and keep his/her eyes open.
- It is important to minimize movements of the TMS coil and the head during assessment of the RMT as even small movements can change the TMS trajectory and influence the measurement.
- If you do not have EMG measuring equipment, the above procedure can be conducted using ***relative visual inspection***. After each pulse, the TMS administrator should visually watch and note whether there was a resulting contraction in the target muscle. If it is too difficult to visualize a muscle twitch in an already contracted muscle, detection of a twitch can be aided by having the participant press the target muscle against another technician to feel the resulting twitch. The lowest intensity that elicits at least 5 out of 10 detectable muscle twitches is the active motor threshold.

3.6 Locating the Phosphene Hotspot

Phosphenes are brief visual percepts that are caused by mechanical or electric stimulation of neurons along the retino-geniculo-striate pathway rather than by photo-transduction. TMS of the visual cortex can activate underlying neurons and pathways to elicit phosphenes that are perceived as small flashes of light or brief scotomas (blurring of a patch of visual field). These flashes can assume varied characteristics depending upon which level of the visual cortex is being stimulated. For instance, stimulation over early visual areas (e.g., primary visual cortex [V1] and secondary visual cortex [V2]) typically elicits stationary phosphenes while stimulation over visual area 5/middle temporal visual processing area [V5/MT+] typically elicits motive phosphenes.

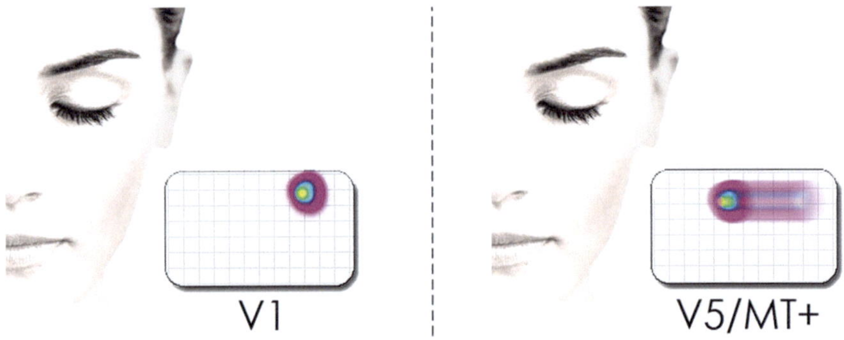

TMS-induced phosphene behavior varies between subjects but is usually consistent within subjects. Although it has been argued that up to 40% of individuals cannot perceive TMS-induced phosphenes, current theories suggest most individuals can perceive phosphenes if they are properly trained on how to do so (to reflect this conception, we will outline a brief "phosphene recognition" protocol below).

Phosphene elicitation can be used for multiple purposes, including to map the functional organization of the visual cortex, to determine state-dependent effects of stimulation, and to explore cortico-cortical interactions underlying visual attention and awareness. Most commonly, however, phosphene elicitation is used to assess the relative excitability of the visual cortex. This measure is known as the phosphene threshold and can be utilized for later determination of procedural stimulation intensity (*Sect. 3.7*).

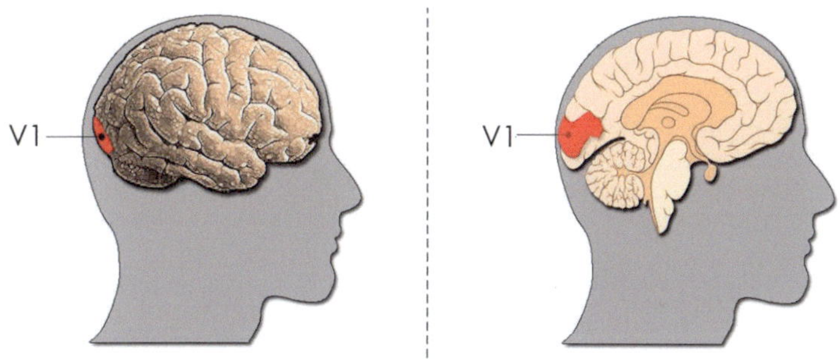

Below, we will determine the left visual field V1 phosphene hotspot.

3.6.1 Steps for Locating V1 Phosphene "Hotspot"

Step 1: Seat the participant comfortably in a chair and place a light-blocking mask over his/her eyes. This will aid in subjective phosphene detection. Avoid having the subject clench their eyes as this can cause mechanical stimulation of the retinal ganglion cells.

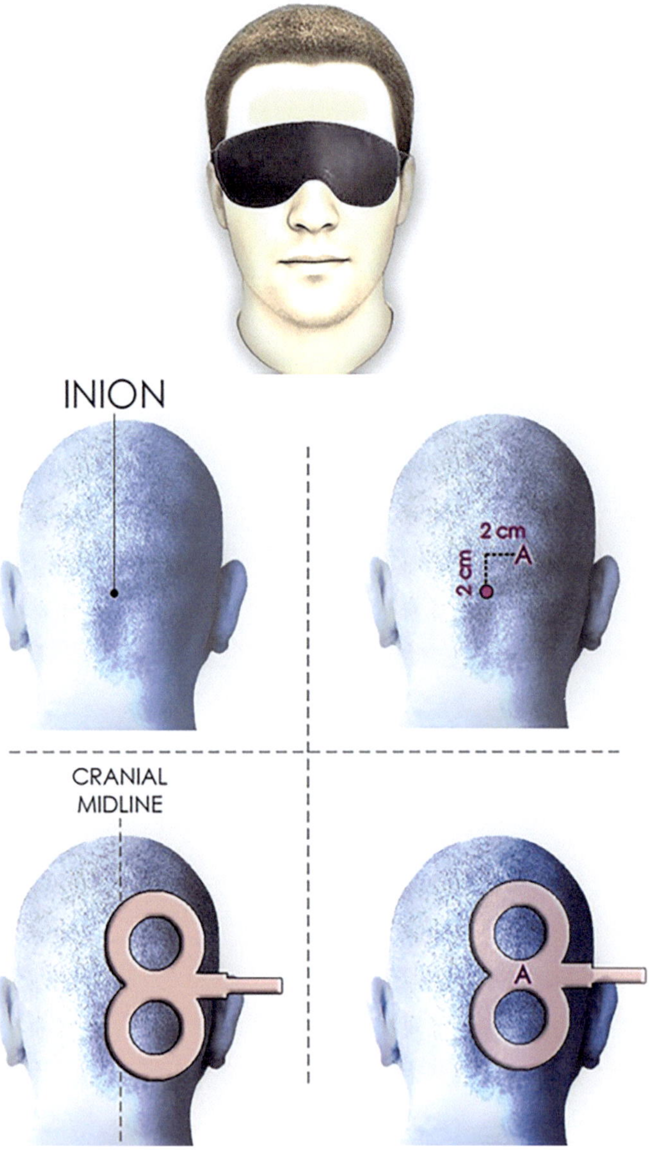

Step 2: Locate the inion. From here, measure 2 cm superior and 2 cm lateral. In this example, this location is marked "A." Hold the TMS coil so the interior curve faces cranial midline. Place the center of the coil over location "A."

Step 3: Set the TMS output to a low intensity, in this example 30% MSO. This is done to acclimate the participant to the sensation without startling them. Give 2–3 pulses (6–10 s apart). After each pulse, ask the subject to report any visual experience in the contralateral visual field.

Step 4: If the participant reports a phosphene, explain that all future phosphenes will have similar characteristics (same size, shape, location, etc.). You may wish to repeat the pulses several times to ensure the participant feels comfortable with the procedure and confident recognizing the phosphene.

Step 5: You can interleave sham pulses (with the coil tilted on its side), or pulses to non-visual areas, to help distinguish between TMS phosphenes and eyeblinks or startle responses.

Step 6: If the subject does not report a phosphene, increase stimulator intensity by 10% MSO and repeat *Step 3* until a phosphene is reported. If the subject still does not report a phosphene at 100% MSO (or the highest intensity that is tolerable), they may be unable to do so; you may wish to excuse them from further participation.

Step 7: Once you reach an intensity that consistently elicits phosphenes at location "A" and the participant can accurately and confidently identify the elicited phosphene, give a single pulse and ask the subject to report any visual experience in the contralateral visual field and rate the vividness of the percept.

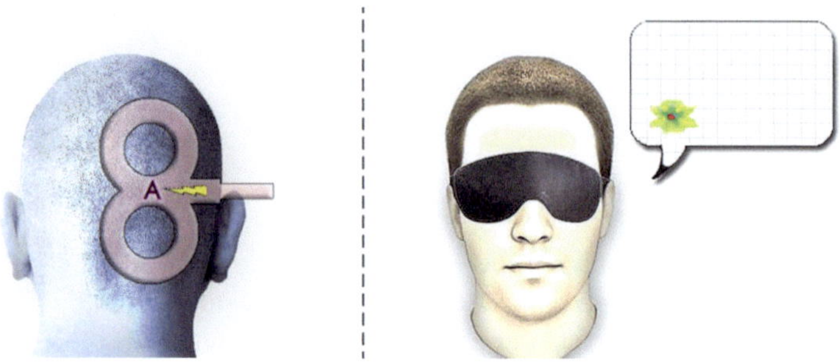

Step 8: Repeat Step 6 at least five times at this location, spacing pulses 6–10 s apart to minimize train effects.

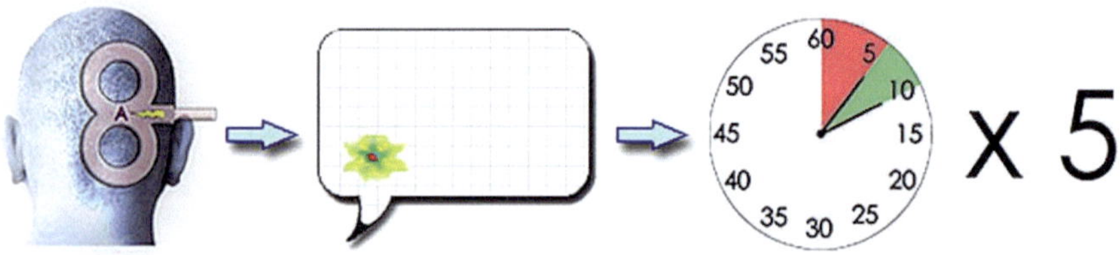

Step 9: Mark a 5 cm by 5 cm grid (in 1 cm intervals) around location "A." Chose a location on this grid and move the center of the TMS coil to this point.

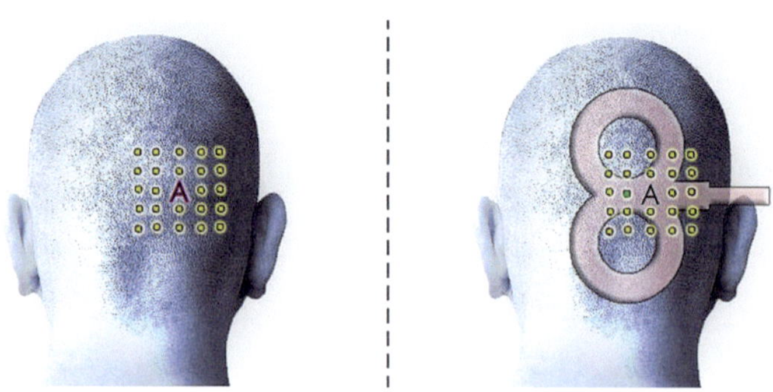

Step 10: Generate a single pulse at this new location, ask the subject to report any visual experience in the contralateral visual field, wait 6–10 s, and repeat at least five times.

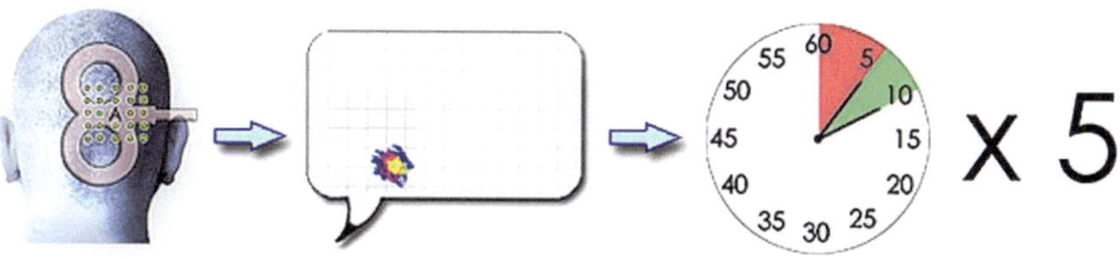

Step 11: Repeat Steps 9 and 10 at various points along the grid and note the phosphene incidence and vividness.

Step 12: Determine which position elicits the most consistent, vivid, and unambiguous subjective phosphene reports. **This position is the V1 phosphene hotspot**.

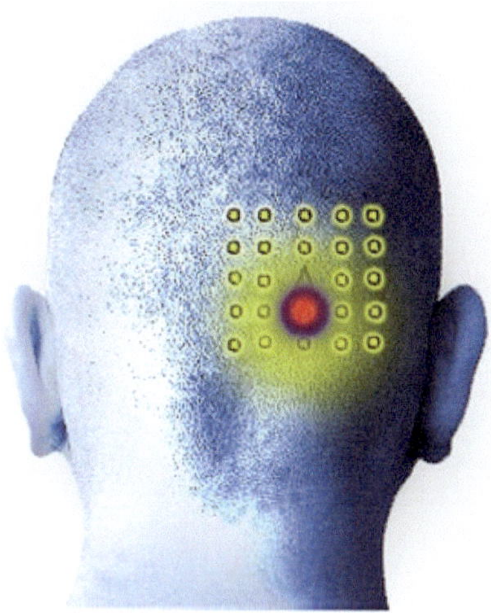

Notes:

- Phosphene hotspots can also be determined for other visual cortices, such as V2 and V5/MT+. Remember phosphene behavior will likely change according to the cortical location stimulated.
- Phosphenes should always appear contralateral to the stimulated hemisphere (except when stimulating the midline). Percepts that are ipsilateral may instead be eyeblinks or other non-specific effects.

- If the participant is having difficulty detecting phosphenes, it may be beneficial to spend a short while allowing him/her to "dark adapt." To do this, simply allow them to rest comfortably with the light-blocking mask in place for approximately 10 min. This procedure may help sensitize the participant to detecting phosphenes.
- If during "*training*" the participant is unable to detect a phosphene, an additional option is to give two pulses immediately back-to-back. This will increase the number of stimulated neurons and should result in phosphenes that appear more vivid and maximize the ability of phosphene detection. If the participant can consistently identify a phosphene using this method, return to single pulses to determine if the participant can maintain accuracy.
- Always hold the coil so that the front faces cranial midline (with the coil handle pointing away from the midline) as this orientation is the most likely to activate visual cortical neurons. The figure below shows *incorrect* coil orientations.

3.7 Determining the Phosphene Threshold

Although TMS intensity during clinical interventions and experimental studies is typically set relative to an individual-specific motor threshold, the visual phosphene threshold can supply a secondary index of cortical sensitivity and excitability in non-motor regions. This measure can be more appropriate to derive intensity for visual and other non-motor cortical stimulation protocols.

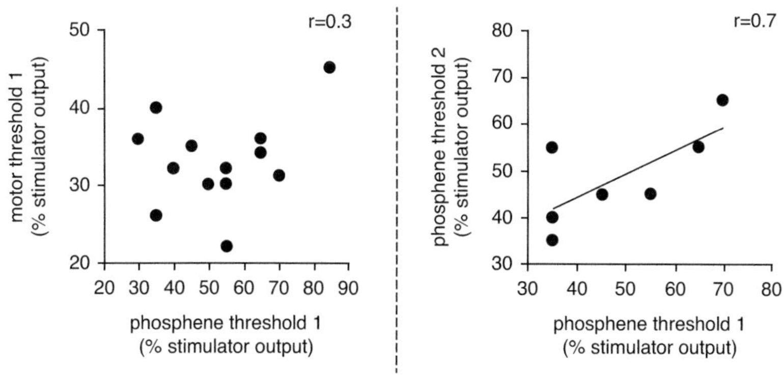

(Stewart et al. 2001)

The phosphene threshold is typically defined as the minimum stimulation intensity required to elicit an unambiguous phosphene in at least half of single TMS pulse trials.

Phosphene threshold, although typically stable within subjects, has been shown to demonstrate greater variability than motor threshold. In addition, phosphene and motor thresholds appear to be mildly correlated, suggesting cortical excitability and reactivity may be globally comparable.

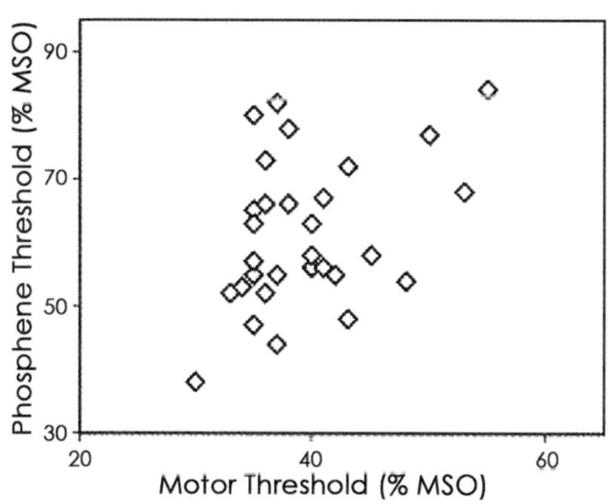

(Gerwig et al. 2003)

Below, we will demonstrate how to determine the phosphene threshold of the left visual field by stimulating V1 of the right occipital cortex.

3.7.1 Steps for Determining the V1 Phosphene Threshold

Step 1: Seat the participant comfortably in a chair and place a light-blocking mask over his/her eyes. This will aid in subjective phosphene detection. Avoid having the subject clench their eyes as this can cause mechanical stimulation of the retinal ganglion cells.

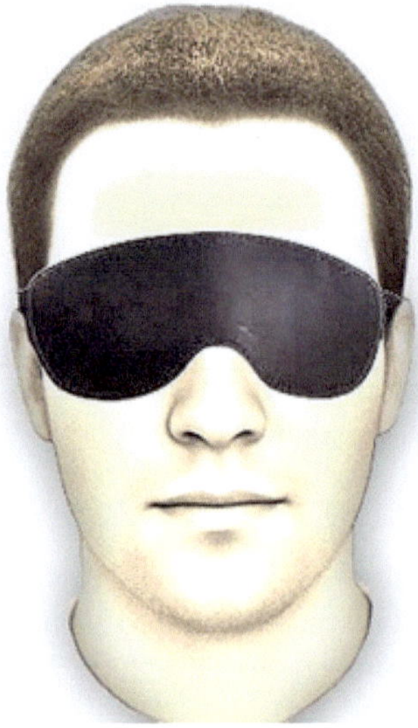

Step 2: Determine right hemisphere V1 phosphene hotspot (*Sect. 3.6*). Hold the coil so that the front faces cranial midline (handle pointing laterally). Remember that when determining the hotspot, one should start at a low intensity (i.e., below threshold) and increase until responses are large and consistently observed (i.e., above threshold).

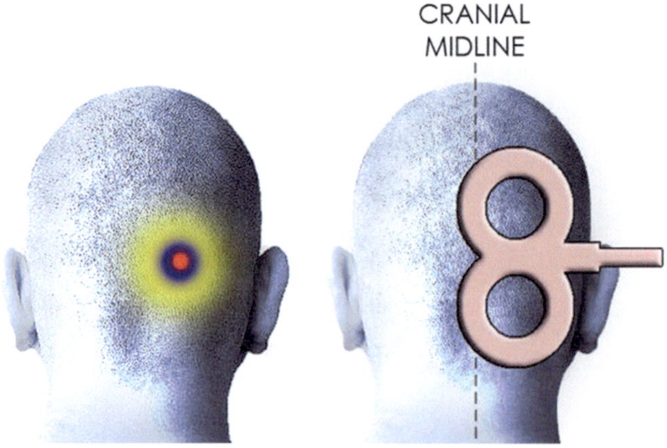

Step 3: Since the intensity reached by the end of *Step 2* is high enough to consistently elicit responses (i.e., above threshold), we expect our phosphene threshold to be lower. In this example (see figure below), the participant reported no phosphenes at 30% MSO, vague or inconsistent phosphenes at 40% MSO, and consistent and vivid phosphenes at 50% MSO.

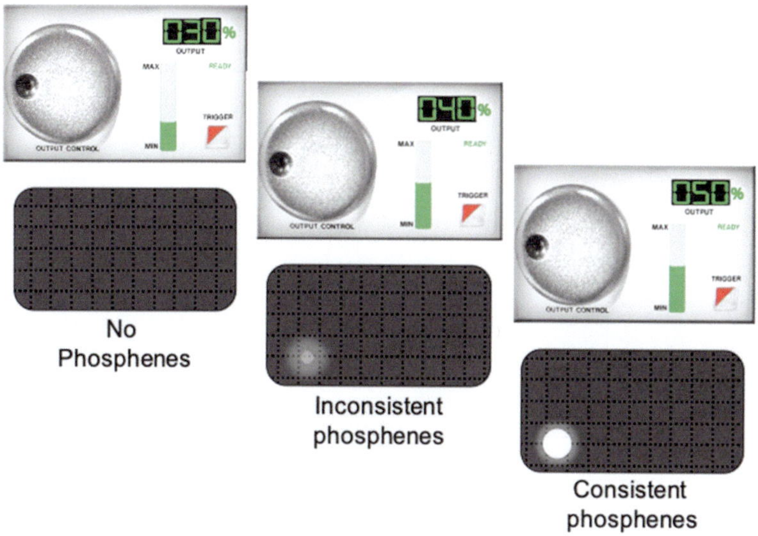

Step 4: Deliver 10 TMS pulses (spaced 6–10 s apart to avoid cumulative effects). For each trial, ask the participant to report whether they perceived a phosphene. Count only "yes" or "no" responses and repeat any trials that the participant is unsure (i.e., "maybe"). In this example, all trials at 50% MSO elicit phosphenes. This indicates that we are *above* threshold.

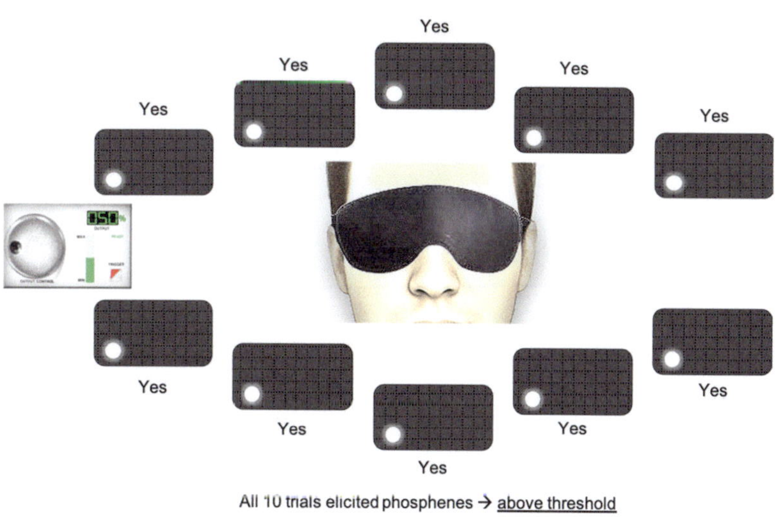

Step 5: If *more* than 5 of 10 TMS trials elicit phosphenes, reduce intensity by 2% MSO and repeat *Step 4*.

Step 6: Repeat *Steps 4–5* until reaching an intensity where *fewer* than 5 of 10 TMS trials elicit phosphenes (in this example, we are at 46% MSO and observe phosphenes in 3 of 10 trials). This indicates that we are *below* threshold.

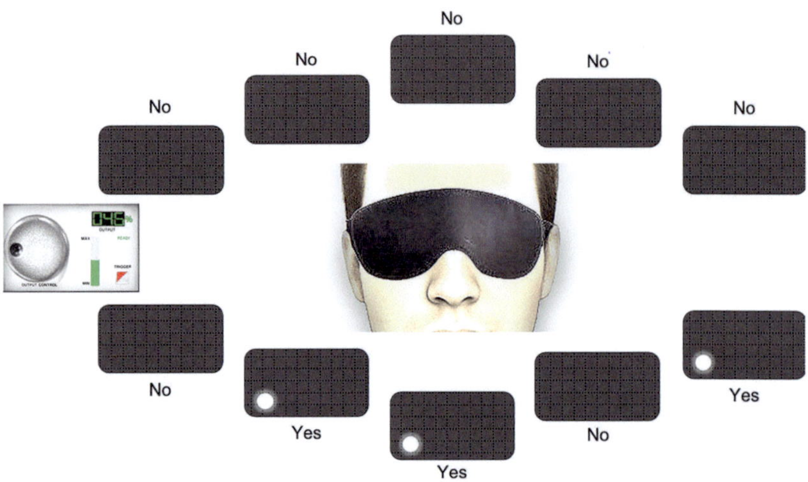

Less than 5 of 10 trials elicited phosphenes → below threshold

Step 7: Increase intensity by 1% MSO and repeat *Step 4*. If at least 5 of 10 TMS trials elicit phosphenes, you have reached the threshold. If not, the threshold is 1% MSO higher (i.e., the last intensity step where at least 5 of 10 trials elicited phosphenes). In this example, **the phosphene threshold is 47% MSO**.

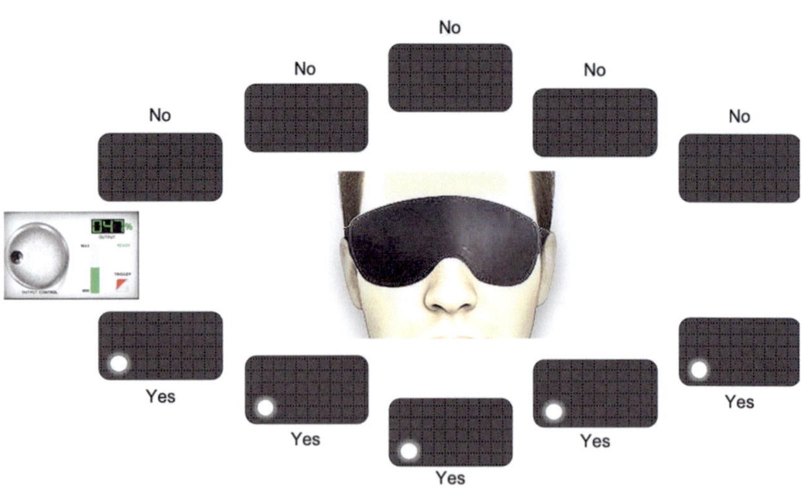

At least 5 of 10 trials elicited phosphenes → at threshold

Notes:

- Phosphene thresholds can also be determined for hotspots in other visual cortices, such as V2 and V5/MT+. Remember phosphene behavior will likely change according to the area of cortex being stimulated.
- If the participant is having difficulty detecting phosphenes, it may be beneficial to spend a short while allowing him/her to "dark adapt." To do this, simply allow them to rest comfortably with the light-blocking mask in place for approximately 10 min. This procedure may help sensitize the participant to detect phosphenes.
- Phosphenes should always appear contralateral to the stimulated hemisphere (except when stimulating the midline). Percepts that are ipsilateral may instead be eyeblinks or other non-specific effects.
- When time constraints are an issue, this procedure can be shortened by using a criterion of three of six phosphenes in Steps 4–7 above (instead of 5 of 10). It is important to note, though, this method has not been formally validated.
- It is important the subject does not clench his/her eyes closed during this procedure as this can cause mechanical activation of retinal ganglion cells leading to phosphene-like percepts.
- It is important to minimize movements of the TMS coil and the head during assessment of the phosphene threshold as even small movements can change the TMS trajectory and influence the measurement.

References

Gerwig, M., Kastrup, O., Meyer, B. U., & Niehaus, L. (2003). Evaluation of cortical excitability by motor and phosphene thresholds in transcranial magnetic stimulation. Journal of the neurological sciences, 215(1), 75-78.

Stewart, L. M., Walsh, V., & Rothwell, J. C. (2001). Motor and phosphene thresholds: a transcranial magnetic stimulation correlation study. Neuropsychologia, 39(4), 415-419.

Further Reading

Section 3.1

Nil

Section 3.2

Nil

Section 3.3

Nil

Section 3.4

Awiszus F (2011). Fast estimation of transcranial magnetic stimulation motor threshold: is it safe?. *Brain Stimul*, 4(1), 58-59.

Di Lazzaro V, Oliviero A, Profice P, Insola A, Mazzone P, Tonali P & Rothwell J C (1999) Effects of voluntary contraction on descending volleys evoked by transcranial electrical stimulation over the motor cortex hand area in conscious humans. *Exp Brain Res* 124, 525–528.

Groppa S, Oliviero A, Eisen A, Quartarone A, Cohen LG, Mall V, et al. (2012). A practical guide to diagnostic transcranial magnetic stimulation: report of an IFCN committee. *Clin Neurophysiol*,123: 858–82.

Rossi S, Hallett M, Rossini PM, & Pascual-Leone A (2009). Safety, ethical considerations, and application guidelines for the use of transcranial magnetic stimulation in clinical practice and research. *Clin Neurophysiol*, 120(12), 2008-2039.

Rossini PM, Barker AT, Berardelli A, Caramia MD, Caruso G, et al. (1994). Non-invasive electrical and magnetic stimulation of the brain, spinal cord and roots: basic principles and procedures for routine clinical application. Reports of the IFCN committee. *Electroenceph Clin Neurophysiol*, 91, 79-92.

Rothwell JC, Hallett M, Berardelli A, Eisen A, Rossini P, Paulus W. Magnetic stimulation: motor evoked potentials. The International Federation of Clinical Neurophysiology. Electroencephalogr Clin Neurophysiol Suppl. 1999;52:97–103.

Silbert BI, Patterson HI, Pevcic DD, Windnagel KA, Thickbroom GW (2013). A comparison of relative-frequency and threshold-hunting methods to determine stimulus intensity in transcranial magnetic stimulation. *Clin Neurophysiol*, 124(4):708-12.

Section 3.5

Nil

Section 3.6

Marg, E., & Rudiak, D. (1994). Phosphenes induced by magnetic stimulation over the occipital brain: description and probable site of stimulation. Optometry & Vision Science, 71(5), 301-311.

Meyer, B. U., Diehl, R., Steinmetz, H., Britton, T. C., & Benecke, R. (1990). Magnetic stimuli applied over motor and visual cortex: influence of coil position and field polarity on motor responses, phosphenes, and eye movements. Electroencephalography and clinical neurophysiology. Supplement, 43, 121-134.

Section 3.7

Antal, A., Nitsche, M. A., Kincses, T. Z., Lampe, C., & Paulus, W. (2004). No correlation between moving phosphene and motor thresholds: a transcranial magnetic stimulation study. Neuroreport, 15(2), 297-302.

Boroojerdi, B., Meister, I. G., Foltys, H., Sparing, R., Cohen, L. G., & Töpper, R. (2002). Visual and motor cortex excitability: a transcranial magnetic stimulation study. Clinical neurophysiology, 113(9), 1501-1504.

Gerwig, M., Kastrup, O., Meyer, B. U., & Niehaus, L. (2003). Evaluation of cortical excitability by motor and phosphene thresholds in transcranial magnetic stimulation. Journal of the neurological sciences, 215(1), 75-78.

Stewart, L. M., Walsh, V., & Rothwell, J. C. (2001). Motor and phosphene thresholds: a transcranial magnetic stimulation correlation study. Neuropsychologia, 39(4), 415-419.

Chapter 4
Basic Neurophysiology Measures

Abstract This chapter describes the rationale and method of several common TMS protocols. The stimulus–response curve evaluates corticospinal excitability, typically using the MEP amplitude across a range of stimulator intensities, from subthreshold to supramaximal, forming a sigmoidal response shape. Thus, the curve captures more information than single-intensity recordings, including motor threshold, maximal amplitude, and a curve slope measure. The cortical silent period (SP) is a phenomenon of electromyographic silence resulting from a TMS pulse during isometric contraction of the target muscle, which is sustained after the MEP, and its length is indicative of cortical inhibition. Central motor conduction time (CMCT) is the time for descending action potentials to arrive at the spinal motoneuron pool following the TMS pulse and is determined by subtracting the peripheral conduction time from the MEP latency. Motor mapping refers to the method of determining a spatial map on the cortex (neuronavigated) or scalp (without neuronavigation), which indicates the boundaries of the target muscle representation, or area, and other quantitative features such as the map center or peak amplitude. The concept of systematic mapping can be applied to the cortical expressive speech area.

D. J. Edwards et al., *A Practical Manual for Transcranial Magnetic Stimulation*,
https://doi.org/10.1007/978-3-031-62304-2_4

4.1 Creating a Stimulus–Response Curve

A stimulus–response curve (also known as *recruitment* curve; *input–output* [*I/O*] curve) measures changes in a response (i.e., outcome variable) by systematically varying stimulus conditions. In this section, we will look at how MEP amplitude changes by varying the stimulation intensity delivered to the cortical hotspot of a target muscle. Sections 5.1 and 5.2 in this manual will use this approach with different stimulation parameters and response variables.

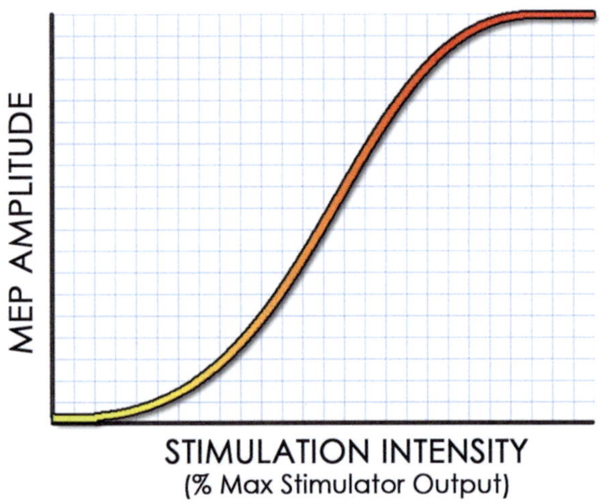

I/O curves are constructed by plotting the muscular response to an incremental increase of TMS intensity. The intensity can be expressed as a percent of the participant's RMT or as fixed intervals of %MSO). Results are typically sigmoidal in shape and can be plotted either linearly (left figure) or using the Boltzmann equation (right figure).

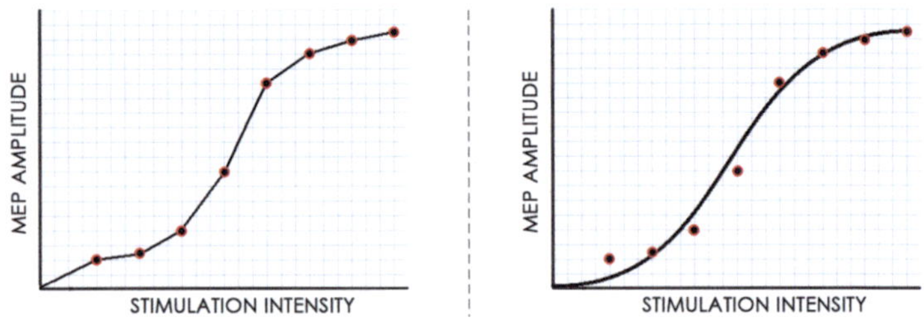

As stimulus intensity increases, it activates neurons with a higher threshold (due to either intrinsic characteristics or spatial distance from the center of the TMS coil). An I/O curve provides a more robust measure of corticospinal projection excitability than single-intensity measurements.

I/O curves can be used to examine changes in corticospinal excitability across disease states, following training regimens, and/or during active drug intake (for instance, I/O slope typically increases with adrenergic activation, and decreases with sodium [Na] and calcium [Ca] channel blockers or with γ-aminobutyric acid [GABA] facilitators).

In this example, we will create an I/O curve for the relaxed right FDI (left M1) using intensity increments of %RMT.

4.1.1 Steps for Creating a Stimulus–Response (I/O) Curve in Left M1 with the Right FDI as the Target Muscle

Step 1: Attach EMG electrodes to the right FDI using a belly-tendon montage with a ground on the bony wrist (*Sect. 3.1*).

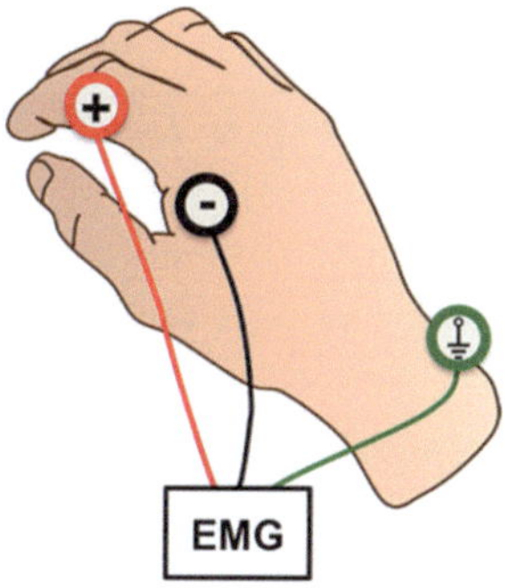

Step 2: Determine the FDI motor hotspot (*Sect. 3.3*). Remember that when determining the hotspot, one should start at a low intensity (i.e., below threshold) and increase until responses are large and consistently observed (i.e., above threshold). The hands should be completely at rest (i.e., no EMG activity) for the hotspot.

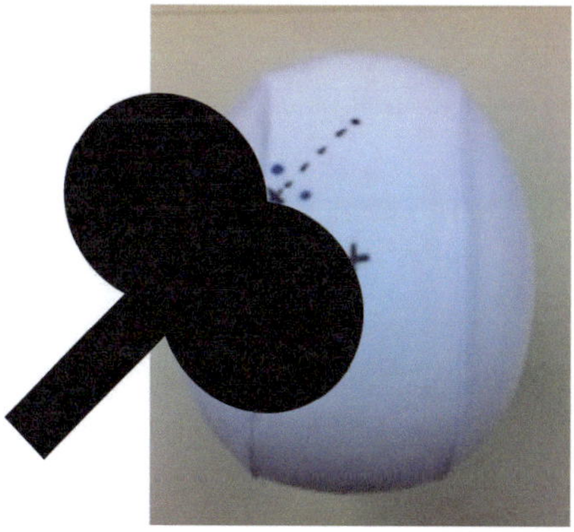

Step 3: Determine the RMT for the FDI hotspot (*Sect. 3.4*). For this example, we will assume an RMT of 50 %MSO.

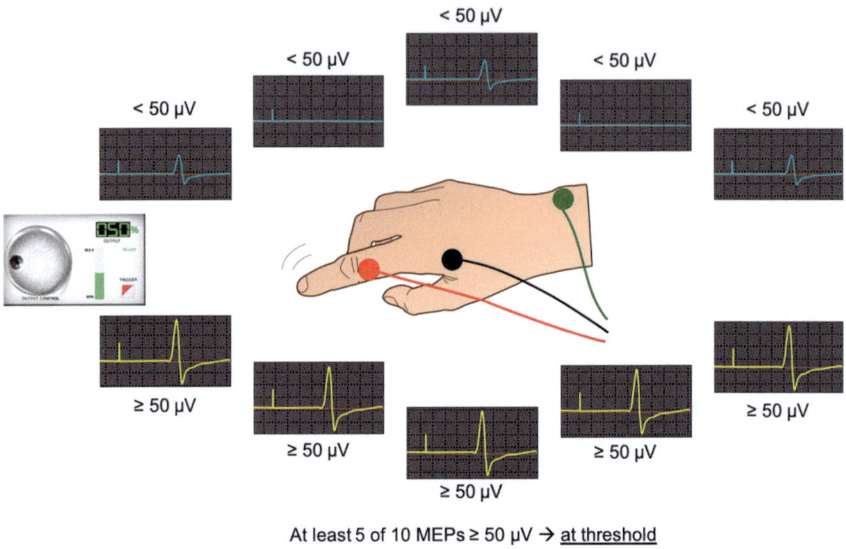

Step 4: Create a table of TMS intensities ranging from 80 to 150 %RMT. We will use this to record the MEP values for each intensity.

%RMT	*80*	*90*	*100*	*110*	*120*	*130*	*140*	*150*
%MSO	40	45	50	55	60	65	70	75
MEP (μV)								

Step 5: Starting with 80 %RMT (40 %MSO in this example), deliver 10 TMS pulses (spaced 6–10 s apart to minimize train effects). Record the resulting MEP peak-to-peak amplitude for each trial. Since we are below RMT, MEPs may not be present and 0 μV values should be included in the average. The I/O curve value for 80 %RMT is the average amplitude for all 10 trials at this intensity.

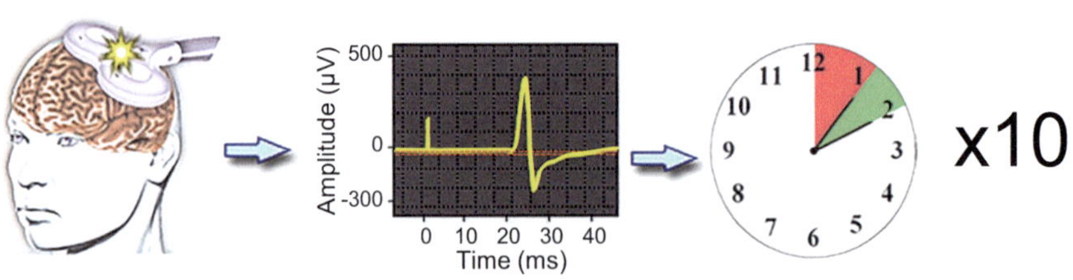

%RMT	*80*	*90*	*100*	*110*	*120*	*130*	*140*	*150*
%MSO	40	45	50	55	60	65	70	75
MEP (μV)	**0**							

Step 6: Increase TMS intensity to 90 %RMT. In this example, this would equal 45 %MSO. Repeat Step 5 and record the average MEP value for 90 %RMT.

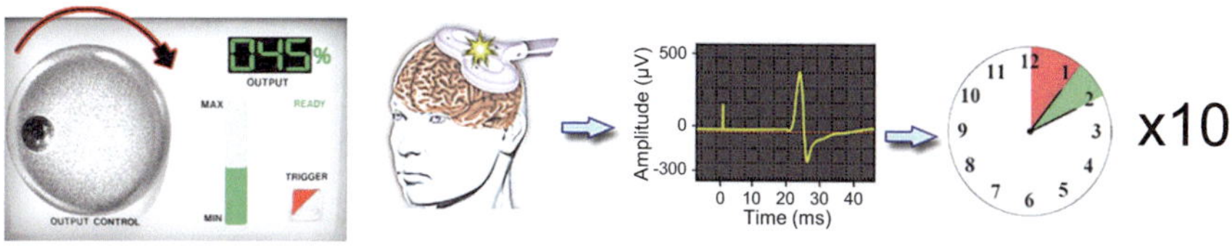

% RMT	*80*	*90*	*100*	*110*	*120*	*130*	*140*	*150*
% MSO	40	45	50	55	60	65	70	75
MEP (μV)	**0**	**10**						

Step 7: Repeat Step 6 for the remaining intensities (i.e., 100, 110, 120, 130, 140, and 150 %RMT). Since we are at or above threshold, we expect MEPs to be present at increasing amplitudes. Note that any trials with no MEP should still be included as 0 μV values in the average. Example values are shown in the table below. Note that actual values will vary from person to person.

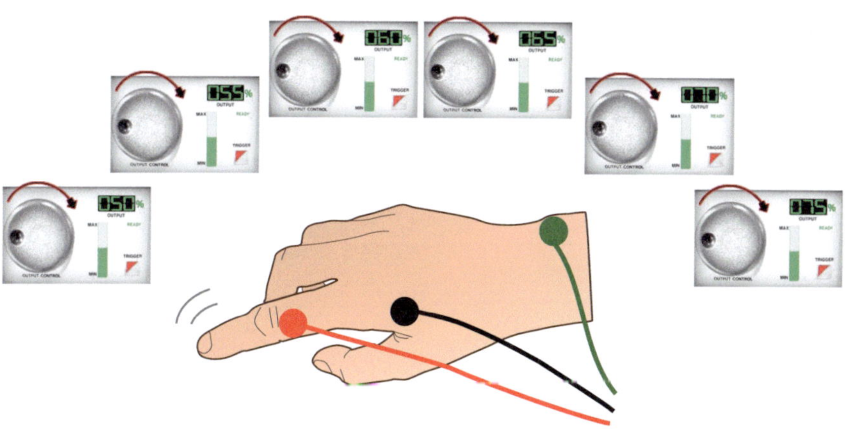

% RMT	*80*	*90*	*100*	*110*	*120*	*130*	*140*	*150*
% MSO	40	45	50	55	60	65	70	75
MEP (μV)	**0**	**10**	**176**	**870**	**1310**	**2512**	**4432**	**4614**

Step 8: Plot the average MEP amplitude (y-axis) as a function of TMS intensity (x-axis). As shown in the figure, it may be possible to reach a plateau where MEP amplitudes do not continue to increase with higher intensities. This is more likely to occur in participants with lower RMT.

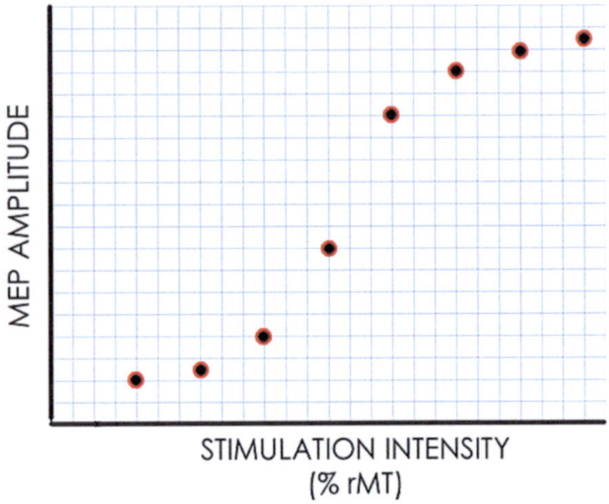

Step 9: Connect the data points or fit the data using a Boltzmann equation.

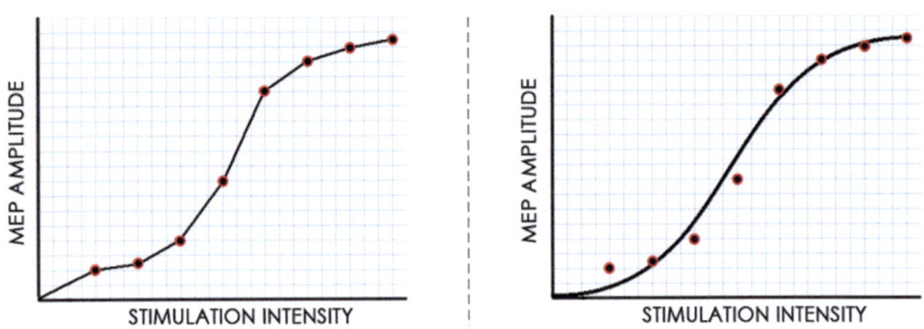

Notes:

- While this example used eight intensity steps from 80 to 150 %RMT (in steps of 10 %RMT), the number of TMS intensity steps and the range between them can be adjusted according to need. Using more intensity steps will lead to a more precise approximation of the I/O curve.
- Note that when intensity steps are based on RMT, it raises the risk that the upper part of the curve will be cut off because the stimulator cannot go beyond 100% MSO. This is more likely to occur in participants with high RMT. An alternative is to set intensity as a % of MSO and normalize to individual RMT in the analysis to compare across individuals or groups.
- This example used intensity steps in increasing increments to simplify the process. However, this can lead to hysteresis. To avoid this, it is recommended that trials are applied at intensities in a pseudorandom order.
- An I/O curve can also be determined using an activated target muscle (see Sect. 3.5). This can be achieved by asking the participant to maintain a minimal activation (~20% of maximal contraction) by pinching the forefinger and thumb together). In this instance, values derived from the RMT should still be utilized (rather than the AMT).

4.2 Assessing the Cortical Silent Period

The cortical silent period (SP) is a measure of intracortical and corticospinal inhibition. A single TMS pulse applied to the motor cortex during tonic contraction of the target muscle elicits an MEP that is larger than if elicited without contraction. This large MEP is followed by a period of suppressed EMG activity that occurs even though the target muscle is still being voluntarily activated. This effect is known as the *cortical silent period* and lasts until the resumption of pre-TMS EMG activity. This period of EMG silence, which may last up to hundreds of milliseconds, can be induced in both the contralateral and ipsilateral muscles. With both approaches, the beginning of the silent period can be either the onset of the MEP (termed *relative* silent period) or the end of the MEP (termed *absolute* silent period). This section will focus on the contralateral cortical silent period (cSP), which is the more common of the two techniques, followed by a brief overview of the ipsilateral cortical silent period (iSP).

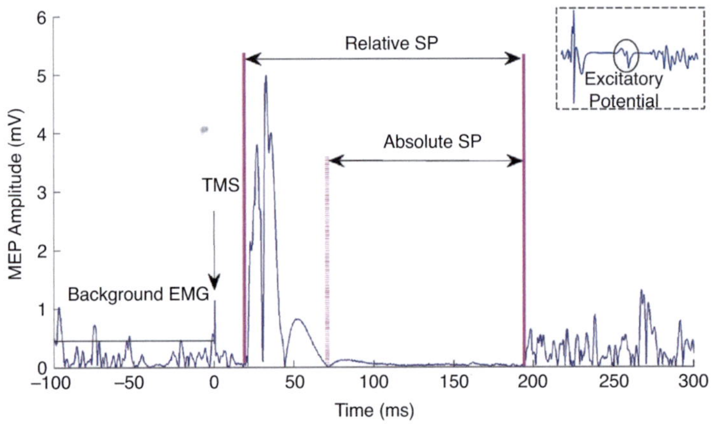

The cSP reflects a combination of cortical activity at the site of stimulation with contributions from the spinal cord. Several lines of evidence suggest that the first ~50 ms of the cSP is due to inhibitory spinal mechanisms, while the remaining portion is mediated by cortical inhibition. Pharmacologic studies have suggested that the cSP is mediated in part by $GABA_B$ receptor neurotransmission. Correspondingly, conditions that are thought to affect $GABA_B$ receptor function or other aspects of cortical excitability can change the cSP duration. For example, the cSP duration is significantly lengthened in paretic limbs following stroke or focal motor epilepsy and the cSP duration can be significantly shortened in limbs affected by Parkinson's disease or the early stages of amyotrophic lateral sclerosis.

The cSP duration can also be influenced by different technical and physiologic factors, including the intensity of TMS and the strength of muscle contraction. The level of TMS intensity and muscle contraction can also affect the presence of a late excitatory potential during the silent period as shown in the figure above. Previous studies have shown that this excitatory potential may increase with a greater degree of voluntary muscle contraction, and it may decrease by increasing the stimulus intensity. It is suggested that this excitatory potential may reflect a period of cortical disinhibition that temporarily permits the reemergence of background EMG activity.

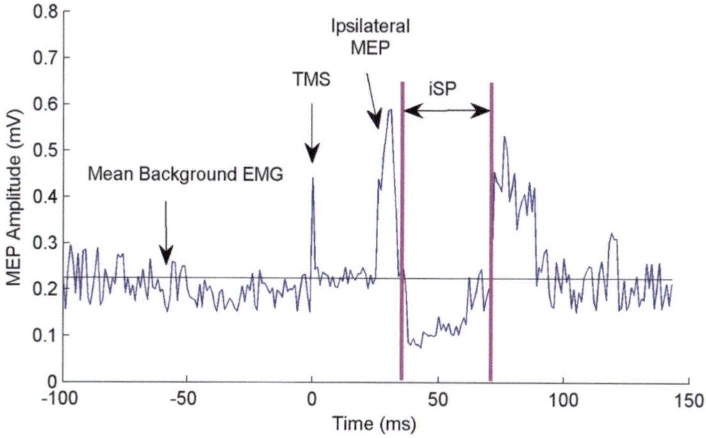

The iSP reflects a transient suppression of EMG activity that occurs when a suprathreshold TMS pulse is applied to M1 during voluntary contraction of the ipsilateral muscle (e.g., contracting the left FDI when targeting the hotspot for the right FDI in left M1). In M1 and many other cortical regions, pyramidal cells in cortical layers II and III send excitatory (glutamatergic) projections along the corpus callosum to the homologous region in the contralateral hemisphere, where they synapse on local inhibitory (GABAergic) circuits. Several lines of evidence suggest that the iSP is mediated by these net-inhibitory transcallosal pathways. This suggestion is based partly on the observation that iSP duration is shortened or absent in patients with lesions of the corpus callosum. Spinal reflexes have been found to remain unchanged during iSP, a finding that further suggests iSP is primarily mediated by supraspinal mechanisms. The iSP duration can be significantly longer in patients suffering from multiple sclerosis and can be significantly shorter in patients suffering from Parkinson's disease. Like the cSP, the iSP duration can be affected by the stimulation intensity and degree of voluntary contraction.

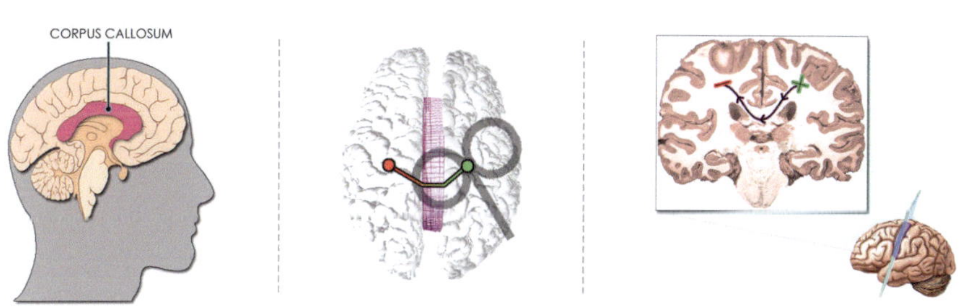

4.2.1 *Steps for Assessing the Contralateral Silent Period in Left M1 with the Right FDI as the Target Muscle*

Step 1: Attach EMG electrodes to the right FDI using a belly-tendon montage with a ground on the bony wrist (*Sect. 3.1*).

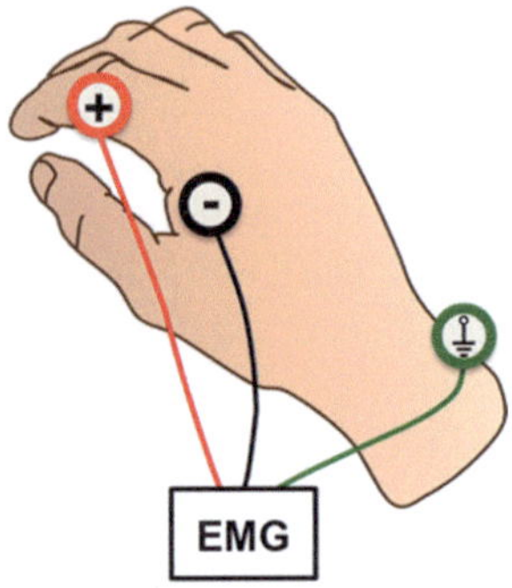

Step 2: Determine the FDI motor hotspot (*Sect. 3.3*). Remember that when determining the hotspot, one should start at a low intensity (i.e., below threshold) and increase until responses are large and consistently observed (i.e., above threshold). The hands should be completely at rest (i.e., no EMG activity) for the hotspot.

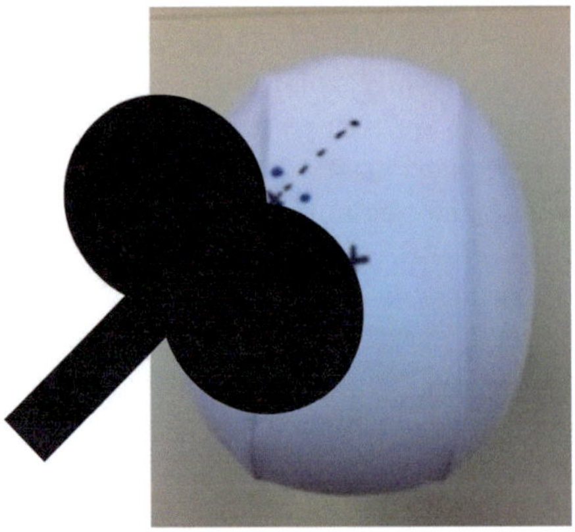

Step 3: Determine the RMT for the FDI hotspot (*Sect. 3.4*). For this example, we will assume an RMT of 50% MSO.

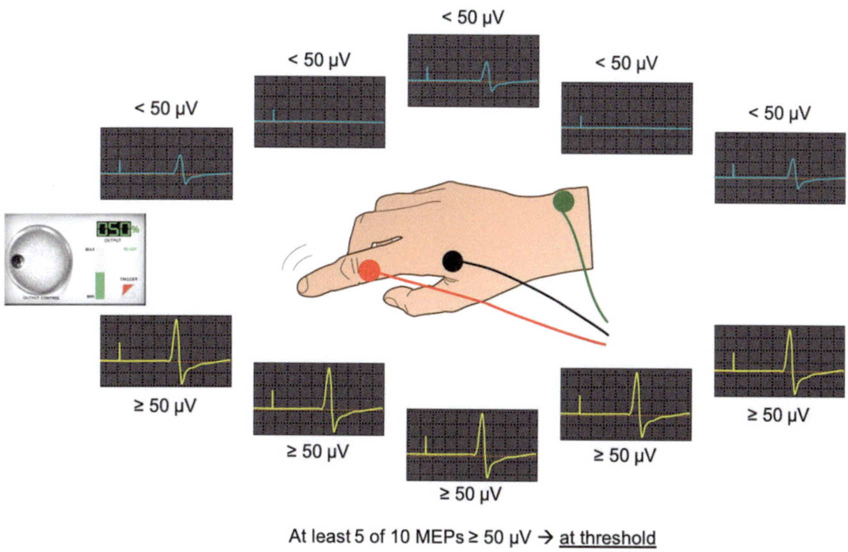

Step 4: Ask the subject to contract the FDI muscle. This can be accomplished by holding the index finger against gravity or by gently squeezing it against the thumb. The target level of activation is about 20% of maximal contraction. If a force transducer is not available, ask the subject to maintain about 100–200 μV of muscle contraction, using the live EMG recording as feedback.

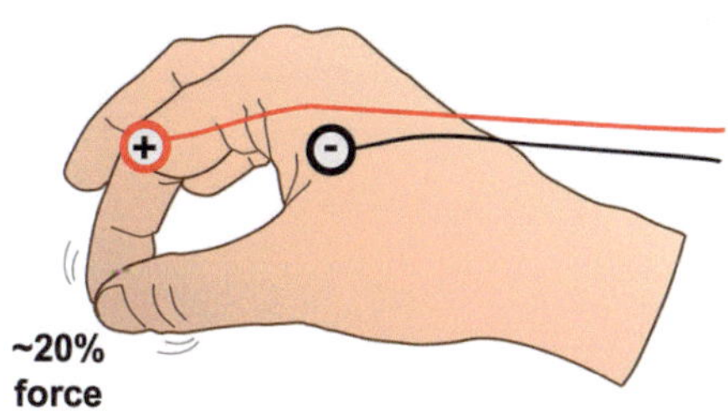

Step 5: Set the TMS intensity to 120% RMT. For this example, this would equal 60% MSO.

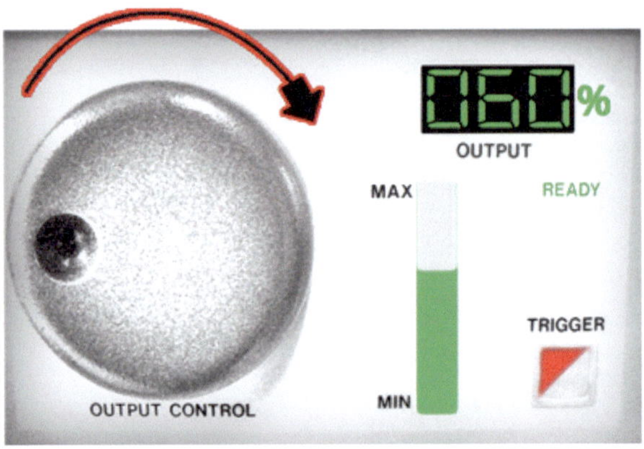

Step 6: Deliver 10 TMS pulses (spaced 6–10 s apart to avoid cummulative effects) to the FDI motor hotspot, while the participant maintains consistent contraction. For each trial, note both the MEP amplitude and the duration of the silent period using either the *absolute* or *relative* measure.

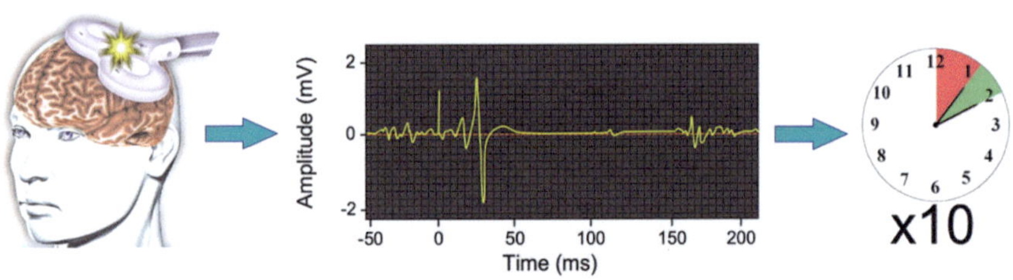

Notes:

– The same protocol can be used to assess the iSP by placing electrodes on both the left and right FDI and having the subject contract the ipsilateral FDI (or both hands simultaneously) when TMS is applied.

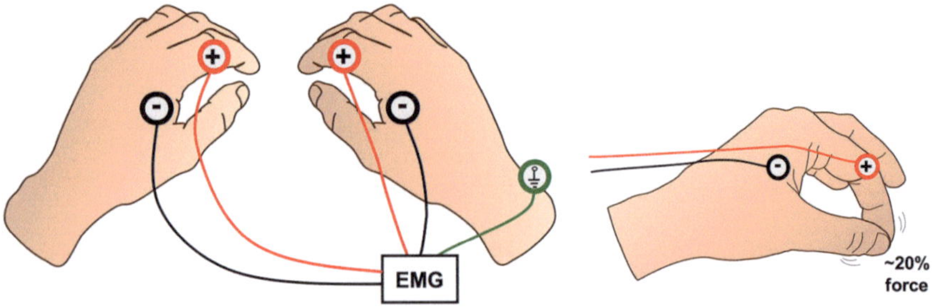

- As the latencies of the ipsilateral MEP and silent period are delayed compared to contralateral ones, the transcallosal conduction time can be determined by subtracting the onset latency of the contralateral MEP from the onset latency of the iSP.
- When using these techniques for clinical diagnostic purposes, it is common to use a higher intensity (~130–140% RMT).
- The cSP (or iSP) can be obtained using AMT (*Sect. 3.5*) instead of RMT (e.g., 150% AMT instead of 120% RMT).
- It may be easier to visualize the beginning and the end of the silent period if EMG traces are rectified by taking the absolute value of the EMG activity.
- In some pathological conditions, e.g., the early stages of amyotrophic lateral sclerosis, it may be difficult to elicit or objectively record an MEP at the beginning of the silent period.
- In addition to silent period duration, peak-to-peak MEP amplitude and the mean level of background EMG activity are often analyzed and reported when conducting the cSP (or iSP) protocol.
- The cSP (or iSP) protocol can be administered as a stimulus–response curve (*Sect. 4.1*) to explore silent period duration as a function of varying stimulator intensity. The slope of this curve has been shown to vary between proximal arm and distal hand muscles.

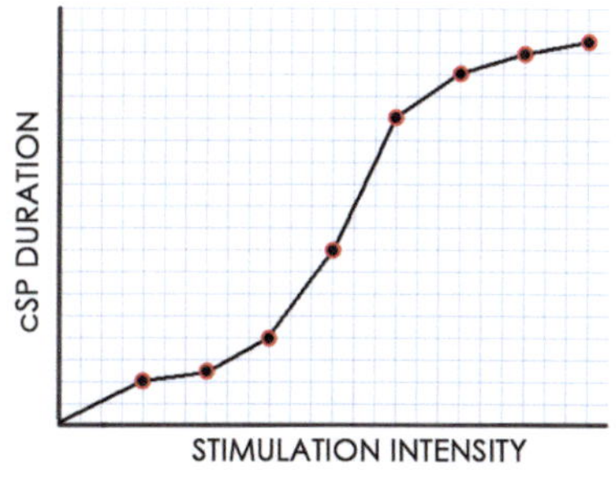

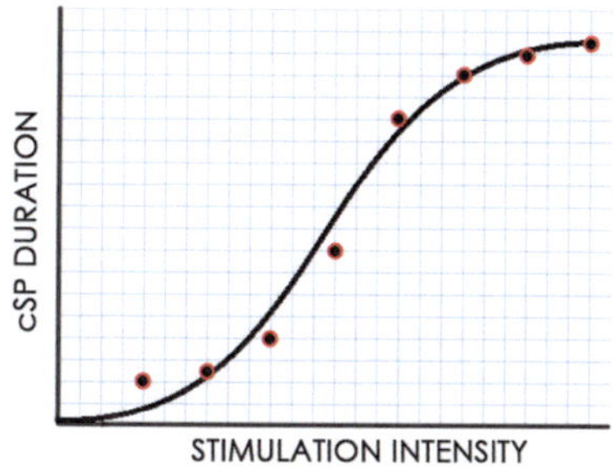

4.3 Central Motor Conduction Time

Central motor conduction time (CMCT) is the time an action potential takes to travel from the primary motor cortex to the spinal cord along the corticospinal tract. The CMCT is the only noninvasive approach to measure the conduction time of the first (upper) motor neuron. CMCT is determined by subtracting the peripheral conduction time (PCT; the time of the second motor neuron from spinal cord to muscle) from the total conduction time (TCT; the time of the first and second motor neurons from M1 to muscle). The TCT is always measured as the latency of an MEP elicited by a TMS pulse over M1. There are two approaches to measure the PCT. This section will focus on using a second TMS pulse at the level of the spine where the nerve roots that innervate the target muscle exit the spinal canal. The other approach, known as the F-wave method, uses an electrical stimulus to the peripheral nerve and will be briefly discussed at the end.

$$\mathbf{TCT}\left(\mathbf{Cortical\,MEP\,Latency}\right) - \mathbf{PCT}\left(\mathbf{Cervical\,MEP\,Latency}\right) = \mathbf{CMCT}$$

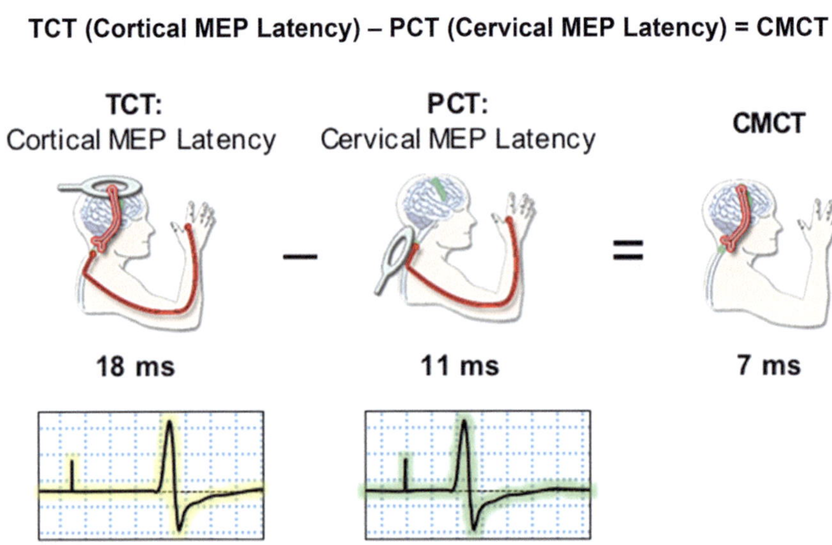

CMCT is the most widely used diagnostic application of TMS. Slower CMCT may indicate several pathophysiologic conditions, including central motor pathway demyelination, loss of the largest and most myelinated axons, and/or dispersion of the descending stimulus volley. Accordingly, slower CMCT is often observed in patients with compressive myelopathy, and in stroke or neurodegenerative diseases (such as multiple sclerosis and amyotrophic lateral sclerosis) when they affect the first motor neuron. Changes in CMCT within a patient can also be used as a prognostic indicator of the disease.

CMCT can be measured in virtually any muscle. It is easiest to measure from intrinsic hand muscles, e.g., the FDI, given the extensive representation of the hand

and fingers in the motor cortex. In the example below, we will use the right FDI as the target muscle to examine CMCT.

CMCT is typically measured with a circular TMS coil, maintaining the same TMS parameters for both cortical and peripheral stimuli. While circular coils are less focal than figure-of-eight coils, they are convenient for stimulating the spinal roots, and placing the circular coil over the vertex/Cz is often sufficient to elicit an MEP.

4.3.1 Steps for Measuring Central Motor Conduction Time in Left M1 with the Right FDI as the Target Muscle

Step 1: Attach EMG electrodes to the right FDI using a belly-tendon montage with a ground on the bony wrist (*Sect. 3.1*).

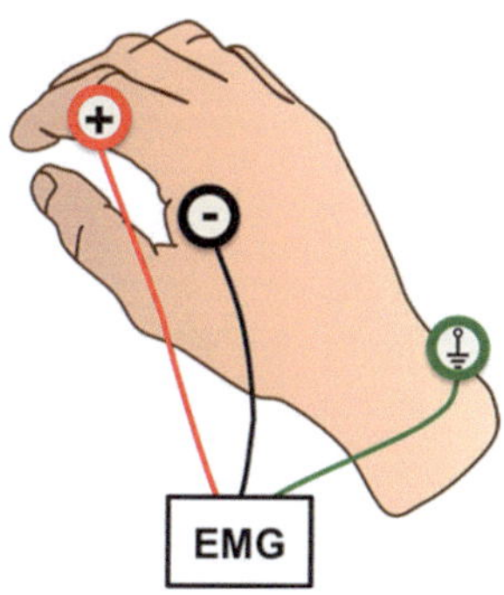

Step 2: Place the circular coil on the head and center it over the vertex (*Sect. 3.2*). If using monophasic pulses (*Sect. 1.5*), the current through the circular coil should be moving counterclockwise so that the direction of the induced current in left M1 is posterior-anterior.

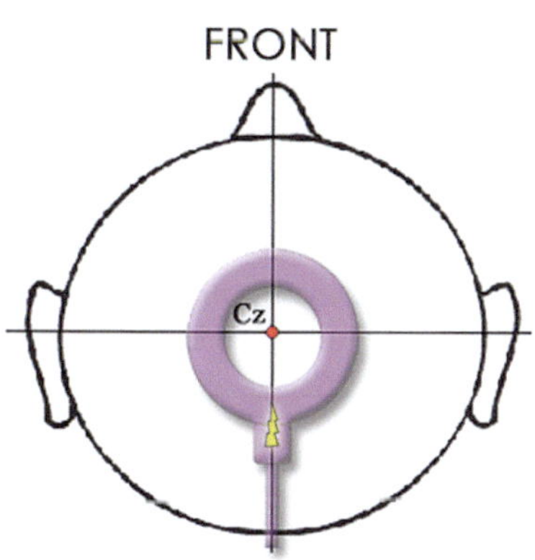

Step 3: When using a 7 cm diameter circular coil centered over Cz, the lateral edge of the coil should be approximately over the motor hotspot. Remember that when determining the hotspot, one should start at a low intensity (i.e., below threshold) and increase until responses are large and consistently observed (i.e., above threshold). If no MEPs are observed at high intensities (~80% MSO), try tilting the coil toward the left (targeted) hemisphere. A systematic search using 1 cm movements (*Sect. 3.3*) can also be used.

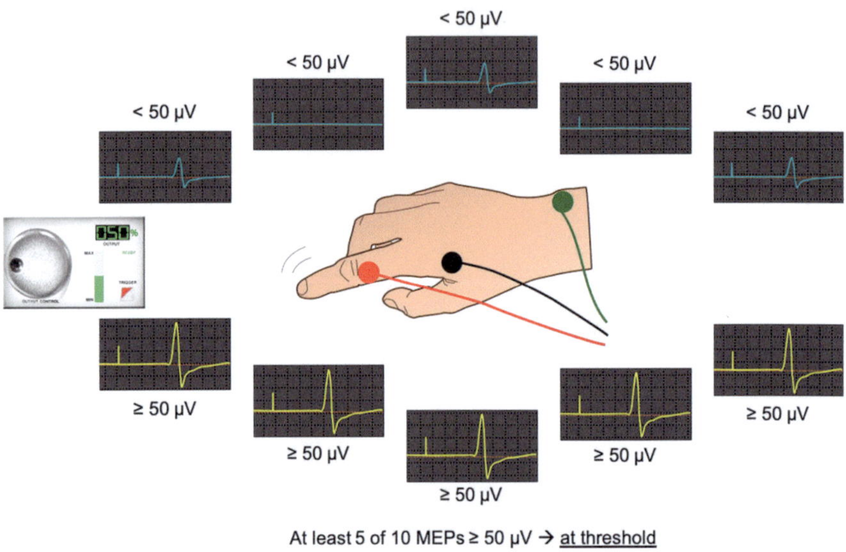

Step 4: Determine the RMT for the FDI hotspot (*Sect. 3.4*). For this example, we will assume an RMT of 50% MSO.

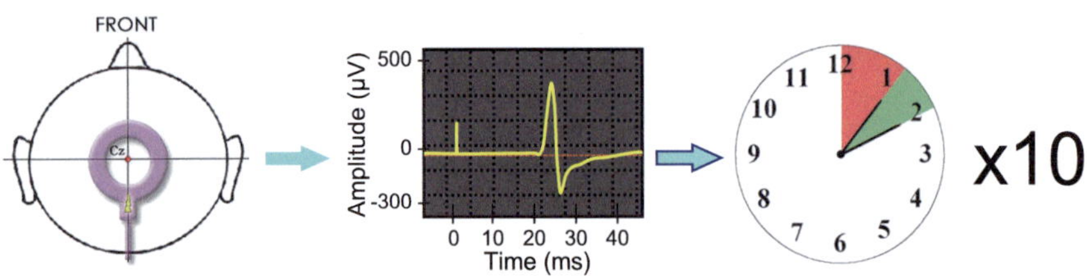

Step 5: Set intensity to 120 %RMT and deliver 10 TMS pulses (spaced 6–10 s apart to avoid cumulative effects) to the FDI motor hotspot. For each trial, note both the amplitude and latency of the MEP (*Sect. 3.3*).

Step 6: Determine the shortest latency to the MEP onset in the series. This "*time to MEP*" value will be labeled **TCT**.

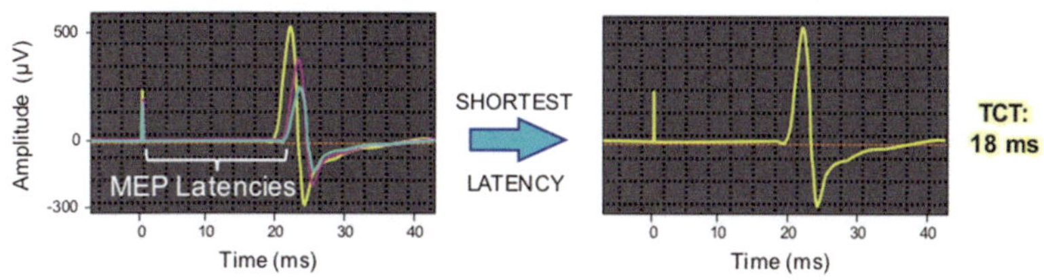

Step 7: Loosen tight clothing around the neck and remove any metallic item in the area, such as necklaces, chains, ID holders, etc.

Step 8: Ask the subject to gently bring down the head and keep it relaxed.

Step 9: Place the coil in a coronal plane over the spinal process of the seventh cervical vertebra (C7; this is the prominent, heavily protruding vertebra at the lower neck/upper back junction). The orientation of the coil does not matter when stimulating spinal roots.

Step 10: Remind the subject to keep their muscles totally relaxed, including the hands, limb, and neck. Explain to the subject that he/she will experience an extension of the head and neck as a result of spinal stimulation and that this may feel uncomfortable.

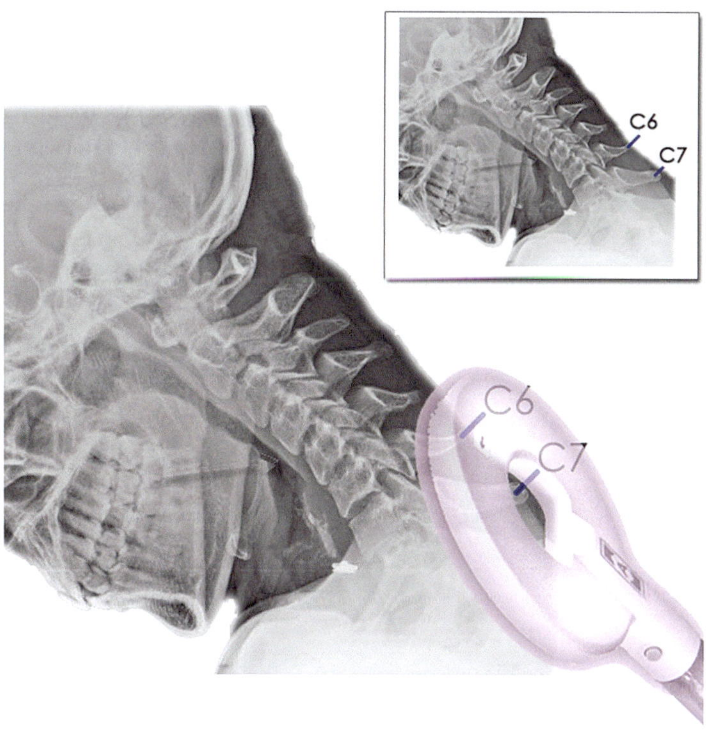

Step 11: Deliver 10 pulses spaced 6–10 s apart using the same intensity as with M1. If the subject complains of discomfort, lower stimulation intensity in 5% steps. If no MEP is elicited, increase stimulation intensity in 5% steps.

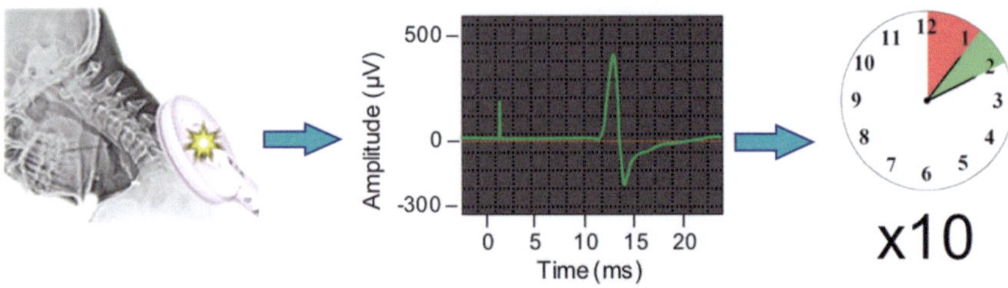

Step 12: Determine the shortest latency to the MEP onset in the series. This "*time to MEP*" value will be labeled **PCT.**

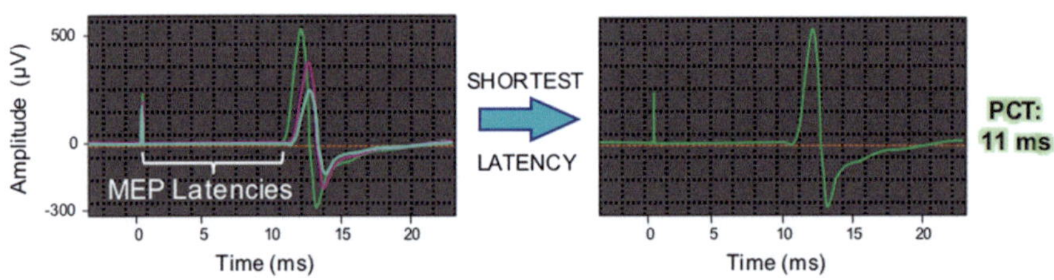

Step 13: Subtract **PCT** from **TCT**. This value is the subject's **CMCT**.

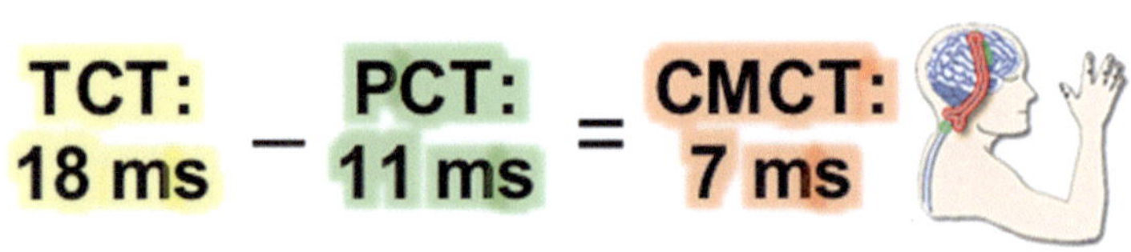

As mentioned in the beginning of this section, an alternative method to measure the PCT is the F-wave method. In this technique, the TCT is calculated the same way as described above.

The PCT is determined by giving a suprathreshold electrical stimulus on the distal portion of a nerve (for FDI the stimulation point is the wrist, proximal to the wrist crease, over the ulnar nerve). This stimulus travels in two directions:

- *Orthodromic (toward the muscle), which elicits an M-wave or muscle response.*
- *Antidromic (back toward the motor neurons' cell body housed in the spinal cord). When the stimulus reaches the spinal cord, a portion of the spinal cord motor neurons will be activated, and the stimulus will travel back to the muscle eliciting an F-wave. The latency of an F-wave indicates the time for the stimulus to travel from the distal stimulating point (wrist) to the spinal cord and back to the muscle (FDI).*

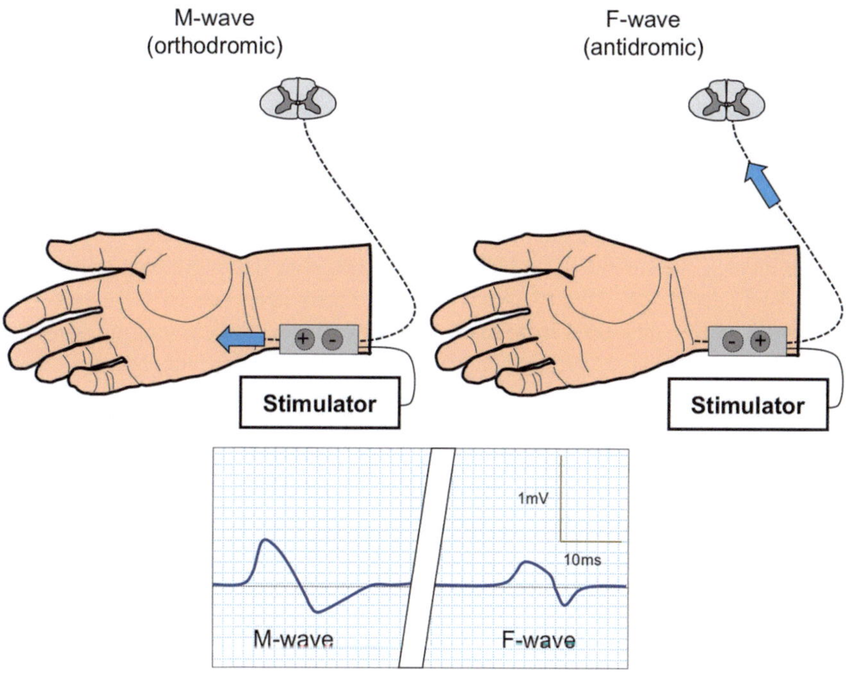

Calculate PCT obtained with the F-wave method using the following formula:

$$PCT = \frac{F\,wave + M\,wave - 1}{2} \qquad CMCT = TCT - PCT$$

Notes:

- There are published normative values for different muscles. Typically, it is only necessary to correct for the height of the individual when assessing CMCT from lower limbs. The standard CMCT value for the FDI should be around 6–8 ms.
- Guidelines (Groppa et al. 2012; Rossini et al. 2015) recommend using an intensity of 120% RMT for research applications and 140% RMT for clinical applications.
- Correct cortical and spinal coil location should elicit nearly identical MEP latency after every pulse. In the instance of widely varying MEP onset delays, you might consider averaging the respective latencies to obtain TCT and PCT (rather than selecting the shortest latency).
- It is important to remember that the different sides of a circular coil conduct current in opposing directions (this is mostly a concern when using a monophasic stimulator, though biphasic pulses often have one dominant phase). The preferential current direction to stimulate the motor cortex is posterior-to-anterior in the cortex (*Sect. 1.5*). Thus, the side of the circular coil with anterior-to-posterior current flow (opposite to that in the brain) will elicit a greater amplitude MEP response in the contralateral hand muscle. Some circular coils are labeled with letters or arrows to indicate the current direction.
- Stimulating the nerve root will elicit a motor response that is peripheral in nature. As such, the shape of the CMAP will be consistent across trials.
- CMCT can be measured from lower limbs by placing the circular coil over the lumbar vertebrae (commonly centered over the first lumbar vertebra; L1); however, a higher intensity may be needed due to the greater depth of the roots and thicker overlying musculature.
- CMCT can be performed with priming of the cortical spinal pathway, either by slightly contracting the target muscle or by imagining muscle contraction. This maneuver may be necessary for patients with impaired motor pathways.

4.4 Motor Mapping

TMS can be used to "map" motor, sensory, and/or cognitive functions onto unique cortical regions via the elicitation of MEPs, phosphenes, or any other measurable effect/behavior.

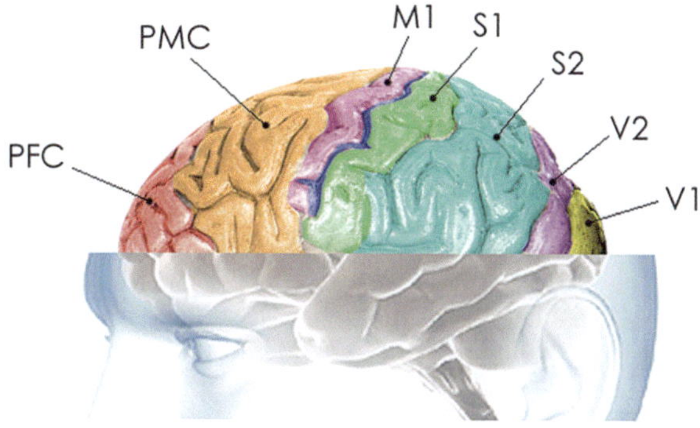

TMS-generated maps are commonly utilized to assess lesion extent, which can be helpful in presurgical planning. TMS mapping requires an MRI-guided coil-tracking system. There are a range of available neuronavigation systems, each with their own instructions. Please refer to those instructions as it is outside the scope of this manual to cover all possible systems. The patient's MRI serves as the anatomical reference upon which the TMS-generated map is constructed. Maps are typically generated by first locating a functional "hotspot" and then stimulating around this location in a coordinated, measured grid. The resultant responses are then analyzed to delineate the boundaries (location and extent) of the map and generate a "topography" of response magnitude.

While many neurophysiological applications of TMS focus on a single muscle of interest, in motor mapping, it is often a goal to map the responses to multiple muscles. While this section focuses on the representation of the FDI muscle, a similar approach can be used to map other muscles of interest, either in serial or in parallel if using a multi-channel EMG system.

When generating a map, there are two important measures to consider: the *hotspot* and the *center of gravity (CoG)*. The hotspot is defined as the location of the largest and most consistent response. In the case of motor maps, the hotspot would be the location that consistently elicits the largest amplitude MEPs and corresponding visible twitch in the target muscle (*Sect. 3.3*). The center of gravity represents the three-dimensional central response location of each unique map. CoG is calculated using MRI coordinates and averaged pulse response. The equation is:

$$X\,Coordinate: \sum(\mathbf{R}_n \mathbf{x} \mathbf{X}_n) / \sum \mathbf{R}_n\; R = Response\,Measurement, X = X\,Coordinate\;and\;n = pulse\,number$$

$$Y\,Coordinate : \sum(\mathbf{R_n\,x\,Y_n}) / \sum \mathbf{R_n}\; R = Response\,Measurement, Y = Y\,Coordinate\,and\,n = pulse\,number$$

$$Z\,Coordinate : \sum(\mathbf{R_n\,x\,Z_n}) / \sum \mathbf{R_n}\; R = Response\,Measurement, Z = Z\,Coordinate\,and\,n = pulse\,number$$

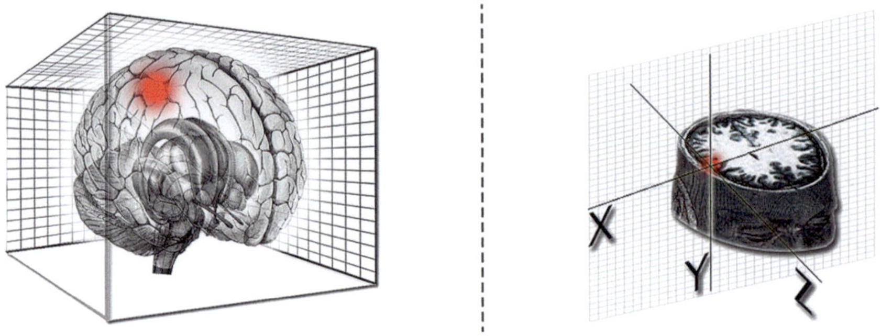

In the case of motor maps, CoG coordinates would be determined by summing the product of each MEP amplitude and its coordinate and dividing by the sum of all MEP amplitudes (for each coordinate X, Y, and Z). In a normal individual, the hotspot should correspond approximately to the determined CoG coordinates; however, lesions or other brain abnormalities can disrupt the topography of motor responses.

In this example, we will generate a motor map for the right FDI muscle as represented in left M1.

4.4.1 *Steps for Creating a Motor Map of Left M1 with the Right FDI as the Target Muscle*

Step 1: Obtain an anatomical MRI of your participant and load it into your preferred neuronavigation (frameless stereotactic) software.

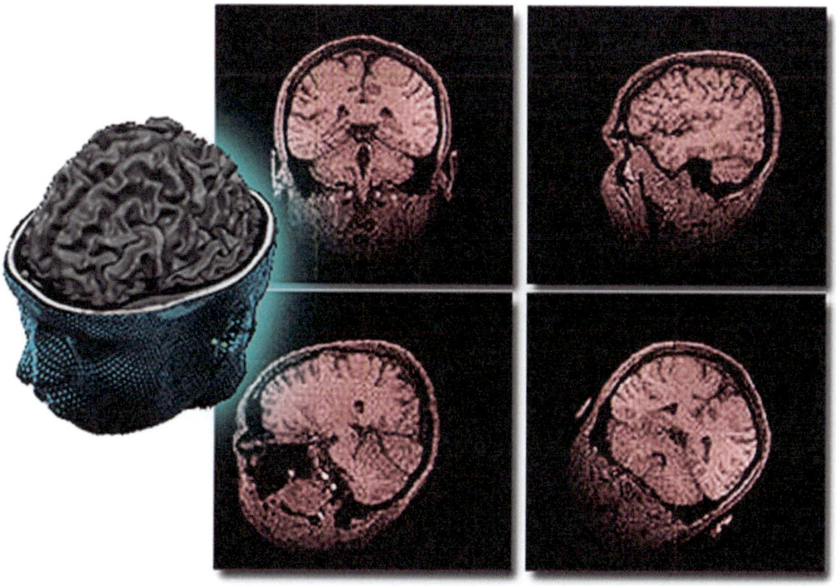

Step 2: Register both the participant's head and the TMS coil with the stereotactic equipment. In this example, we are using infrared-based optical trackers.

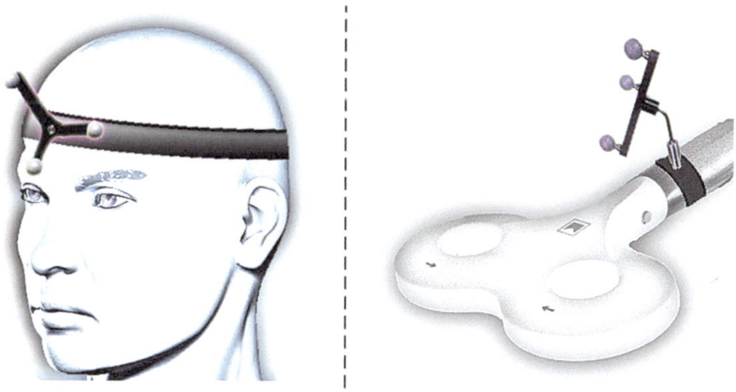

Step 3: Attach EMG electrodes to the right FDI using a belly-tendon montage with a ground on the bony wrist (*Sect. 3.1*)

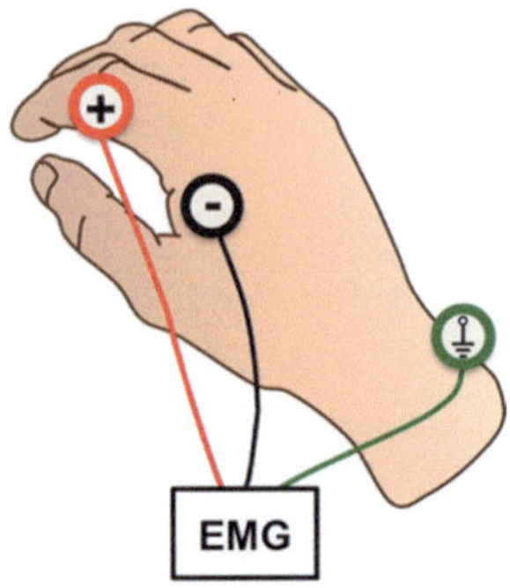

Step 4: Determine the FDI motor hotspot (*Sect. 3.3*). Remember that when determining the hotspot, one should start at a low intensity (i.e., below threshold) and increase until responses are large and consistently observed (i.e., above threshold). The hands should be completely at rest (i.e., no EMG activity) for the hotspot.

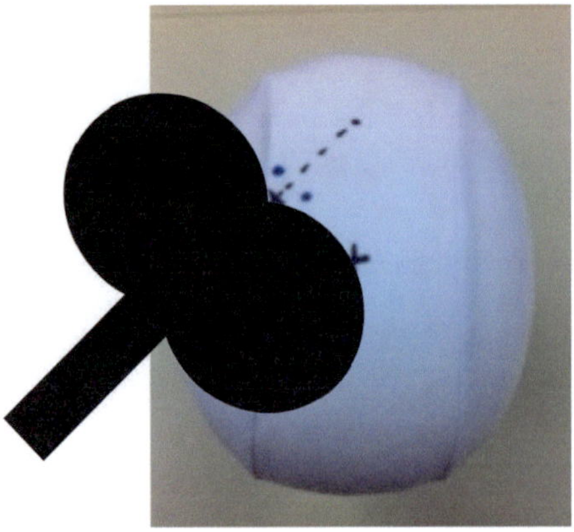

Step 5: Determine the RMT for the FDI hotspot (*Sect. 3.4*). For this example, we will assume an RMT of 50% MSO.

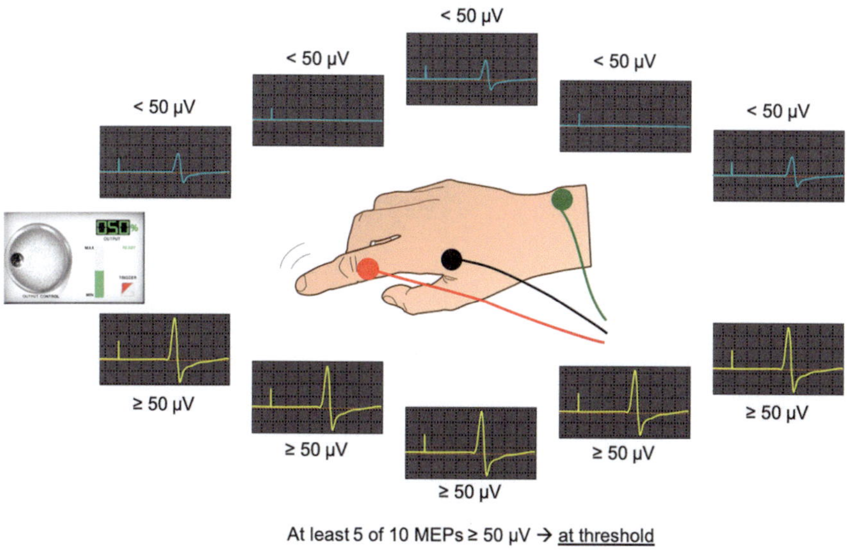

Step 6: Set TMS intensity to 110% RMT. For this example, this would equal 55% MSO.

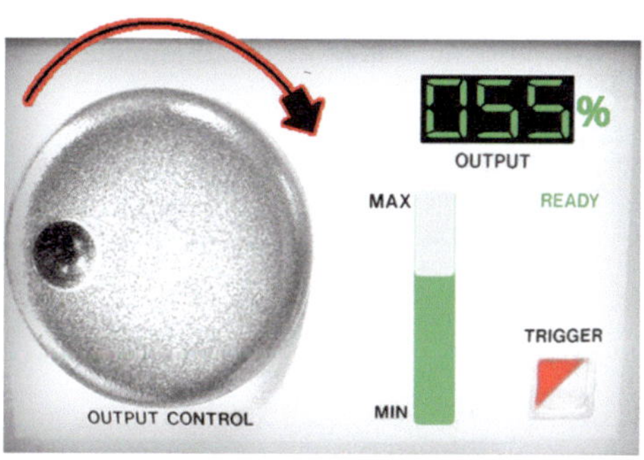

Step 7: Create a 5 cm by 5 cm grid centered on the FDI hotspot, marking points at 1 cm intervals (some neuronavigation systems will do this for you on the subject's MRI).

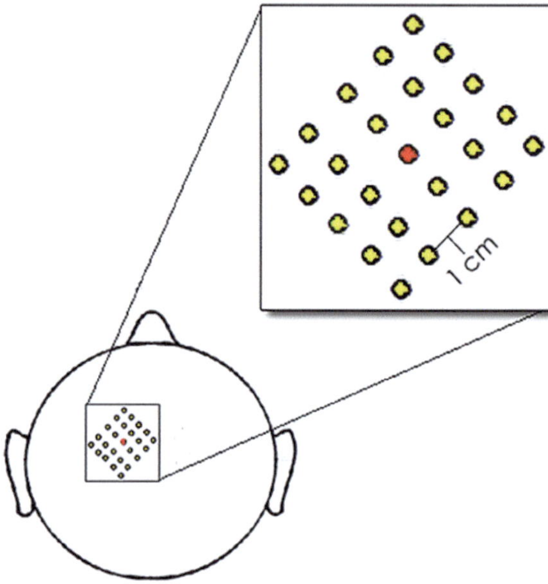

Step 8: Starting in the center of the grid and working outward in a spiral pattern, generate a single TMS pulse at each point. Maintain the same coil orientation as for the hotspot (e.g., 45° from the mid-sagittal plane).

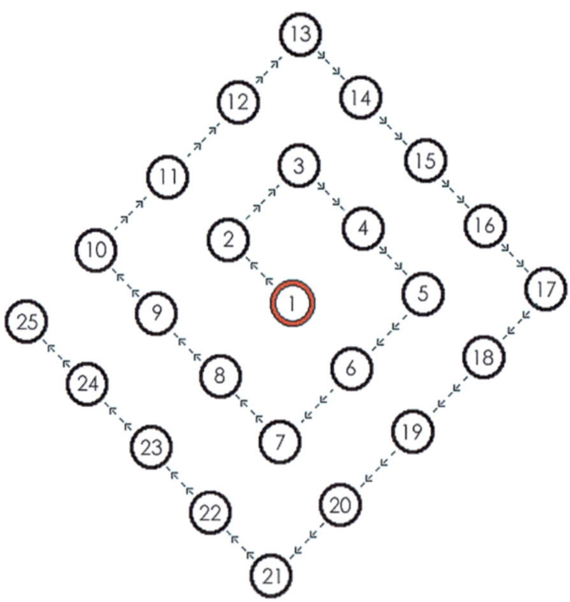

Step 9: Each pulse will have two corresponding measurements: the MEP peak-to-peak amplitude (including 0 if no MEP is present) and the coordinates (X, Y, Z) as given by the neuronavigation system. Record these two measurements.

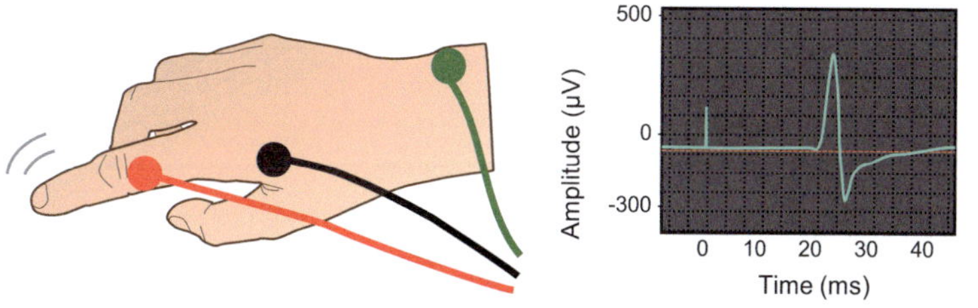

Step 10: To visualize the magnitude of MEP responses in relation to the anatomical location, create a graph and/or topographic color chart of the MEP amplitude. You can visualize the map independently or as an overlay on the MRI for localization purposes.

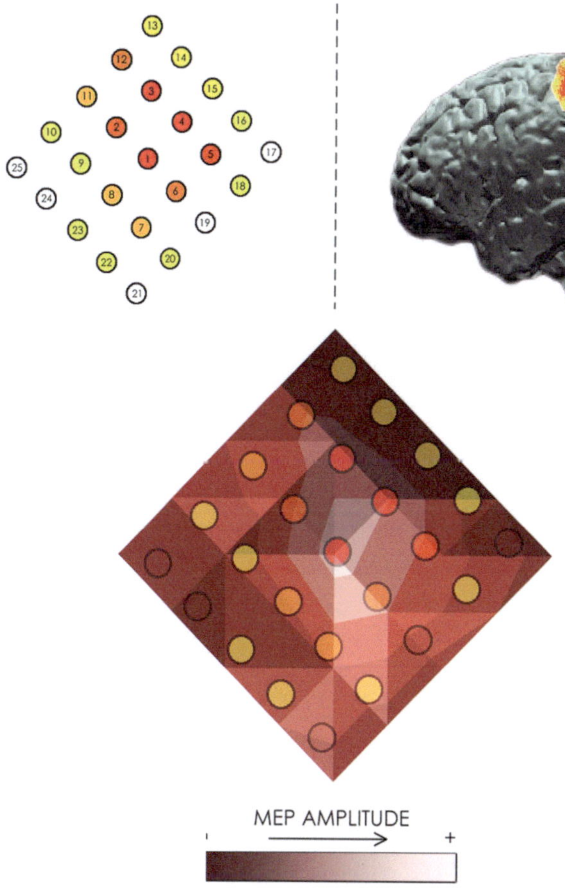

Notes:

- Identifying the hotspot can be challenging due to the presence of a brain abnormality. In some cases, there may be multiple "hotpots," in others, none. Alternatively, you can center the grid on the hand "knob" or other anatomical target on the MRI.
- You may need to adjust the intensity and/or the dimensions of the map. For example, if large MEPs are present in the outermost grid points, reduce stimulation intensity to 105% RMT and recalculate map.
- Brain abnormalities may disrupt the preferred orientation of the neural components to TMS. For sites near important regions (e.g., brain lesions) or when responses are unclear, increase the number of TMS pulses and collect responses with varying coil orientations. When creating the map, the MEP response for a given site can be the largest response recorded for that site, or the average amplitude.
- Mapping can be performed in non-motor regions when there is a sufficient measure of the response to a TMS pulse (e.g., a phosphene; *Sect. 3.6*). Ideally, the responses should have a graded "strength" rating rather than a dichotomous "*yes/no*" response. Language mapping will be discussed in *Sect. 4.5*.

4.5 Language Mapping

Repetitive transcranial magnetic stimulation (rTMS) can be used to transiently disrupt functional processing within a specified neural region (*Chap. 6*). When delivered "online," i.e., during an active task or behavior, this disruption can manifest as an impairment or cessation of the behavior.

Broca's area, located in the vicinity of the left peri-sylvian cortex, is a brain region considered critical for speech production. The scalp location is between EEG coordinates F3 and F7 according to the International 10-20 System for EEG electrode placement, corresponding to the pars triangularis (Brodmann area 45), and the pars opercularis (Brodmann area 45). High-frequency rTMS trains applied to Broca's area during active speech production typically causes a slowing or distortion of vocalizations. Known as overt speech arrest, this phenomenon lasts for the duration of stimulation and speech immediately returns to normal following rTMS cessation.

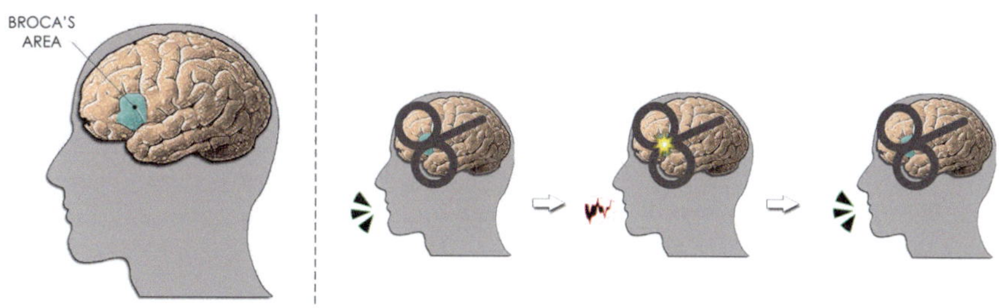

Although some believe this form of speech arrest is muscular in nature, several experiments have demonstrated a hemispheric dominance suggesting the arrest does, indeed, occur due to disruption in cortical linguistic processing. In addition, a purely motor speech arrest can be produced by stimulating the motor cortex, approximately 3 cm caudal to the peri-sylvian region. This dissociation is further suggestive of purely linguistic speech arrest produced with stimulation over Broca's area.

Interestingly, high-frequency rTMS trains applied over the left inferior frontal gyrus (near EEG coordinate F7) during *internal* speech production have been demonstrated to cause a slowing of internal speech (internal speech being unspoken, internal dialogue). Known as covert speech arrest, this phenomenon also lasts the duration of stimulation and internal speech returns to normal following rTMS cessation. As before, covert speech arrest has been demonstrated to show a hemispheric dominance and strong frontal localization suggesting it is cortical (rather than muscular) in nature.

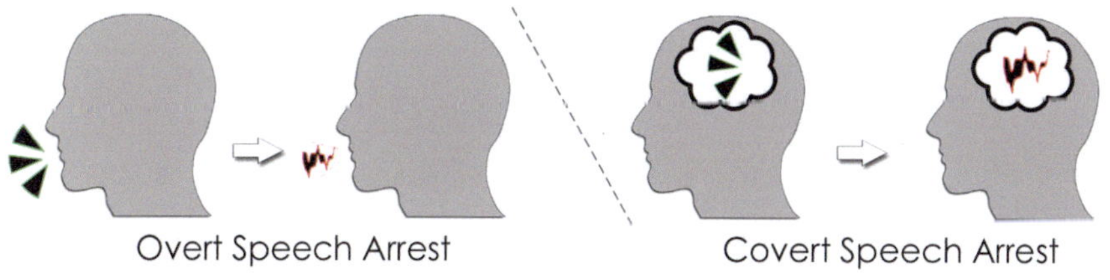

Speech arrest can be combined with mapping techniques (*Sect. 4.4*) to experimentally or therapeutically map speech functions to specific cortical regions. This type of mapping can be used to augment or replace the invasive Wada test currently used to map linguistic function prior to ablative surgery for epilepsy.

Below, we will look at how to induce both overt (aloud) and covert (internal) speech arrest.

4.5.1 Steps for Inducing Overt Speech Arrest

Step 1: Attach EMG electrodes to the right FDI using a belly-tendon montage with a ground on the bony wrist (*Sect. 3.1*).

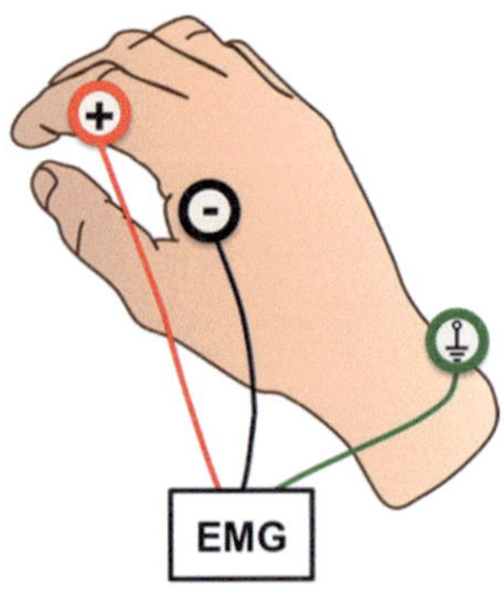

Step 2: Determine the FDI motor hotspot (*Sect. 3.3*). Remember that when determining the hotspot, one should start at a low intensity (i.e., below threshold) and increase until responses are large and consistently observed (i.e., above threshold). The hands should be completely at rest (i.e., no EMG activity) for the hotspot.

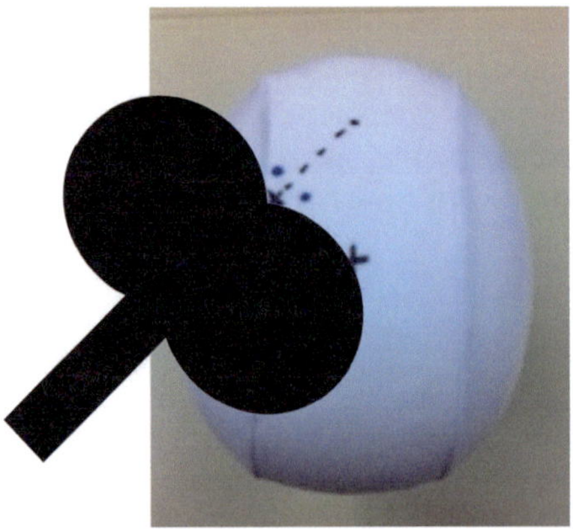

Step 3: Determine the RMT for the FDI hotspot (*Sect. 3.4*). For this example, we will assume an RMT of 50% MSO.

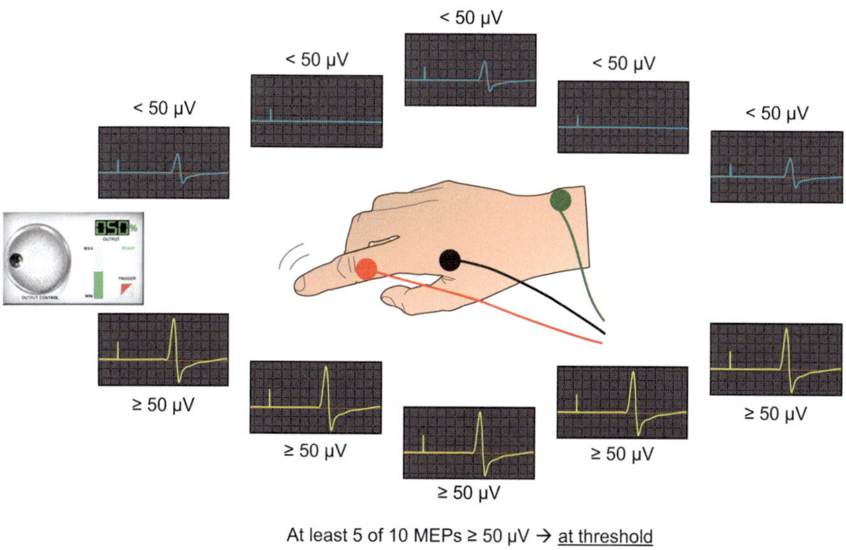

Step 4: Set TMS intensity to 100% RMT. For this example, this would equal 50% MSO.

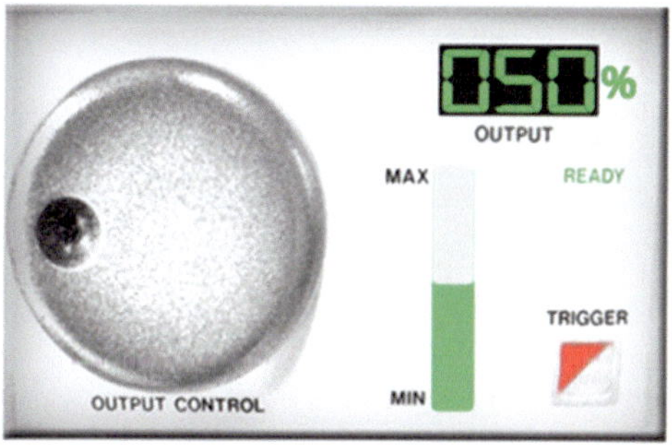

Step 5: From the right FDI hotspot, measure approximately 5 cm anterior and 2.5 cm lateral (toward the cheekbone). This site falls between EEG coordinates F3 and F7 and roughly corresponds to the upper portion of the temple. Place the center of the coil on this site with the handle pointing toward the back of the head to minimize activation of sensitive facial muscles.

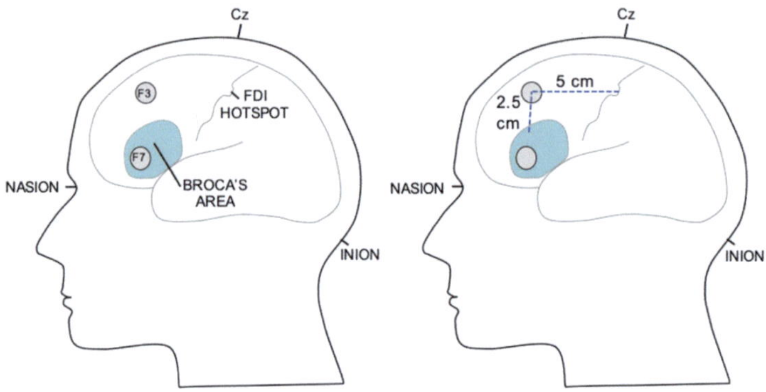

Step 6: Create a 5 × 5 cm grid centered on the site in Step 5, marking points at 1 cm intervals (alternatively, a 3 × 3 cm grid in 0.5 cm increments can be used). If using a neuronavigation system to perform the mapping (*Sect. 4.4*), mark these locations on the subject's MRI.

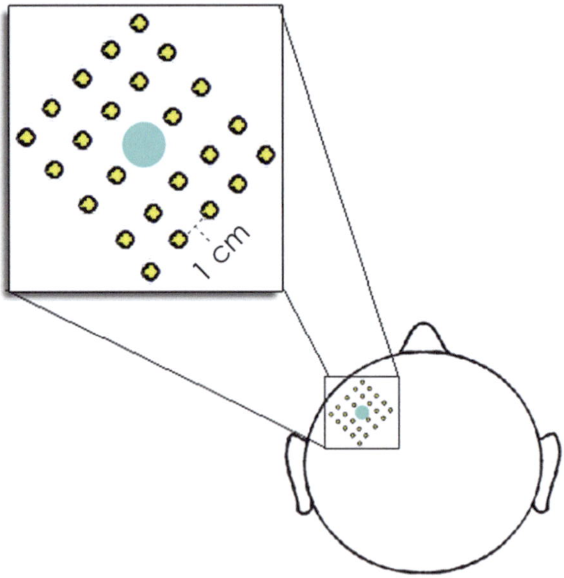

Step 7: Ask the participant to begin counting from zero. Instruct them to speak in a clear, distinct voice that is loud enough to hear over the sound of TMS.

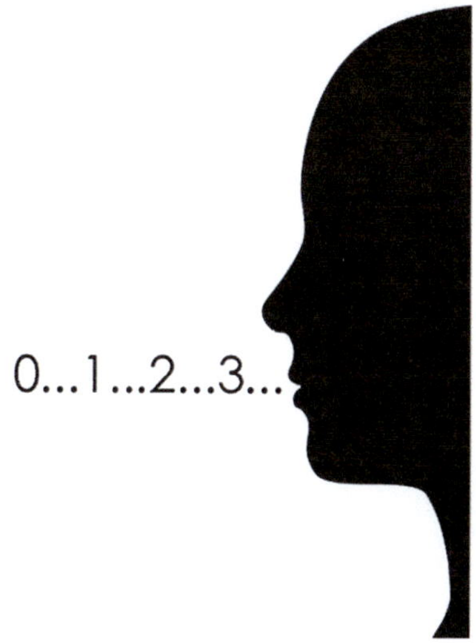

Step 8: Starting in the center of the grid, apply a 5 Hz rTMS train (200 ms inter-pulse interval) for 5–10 s (25–50 pulses total). Note if speech is disrupted during stimulation and rate the severity of the disruption.

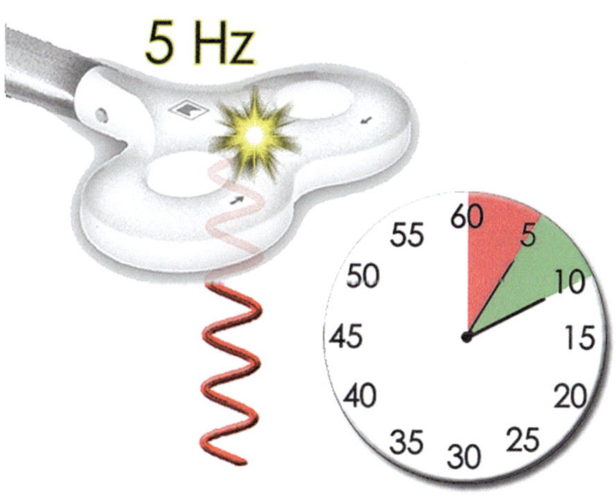

Step 9: Repeat Steps 7 and 8 for the remaining locations on the grid, working out from the center in a clockwise spiral pattern. Note the occurrence and severity of speech arrest.

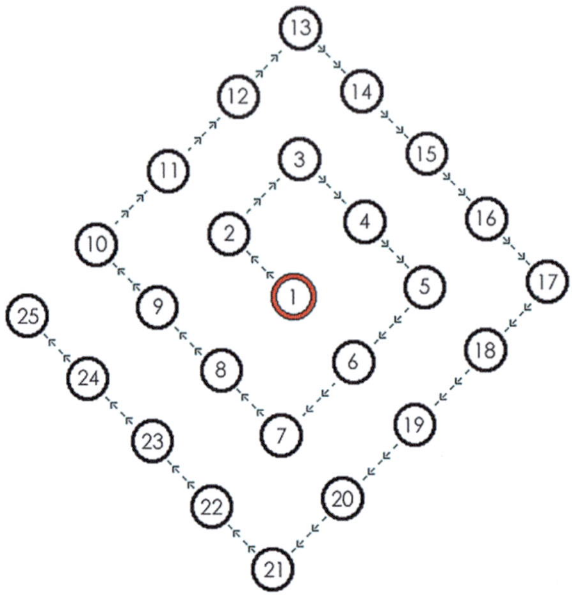

Step 10: Continue this systematic search until a region of noticeable and replicable speech arrest is located.

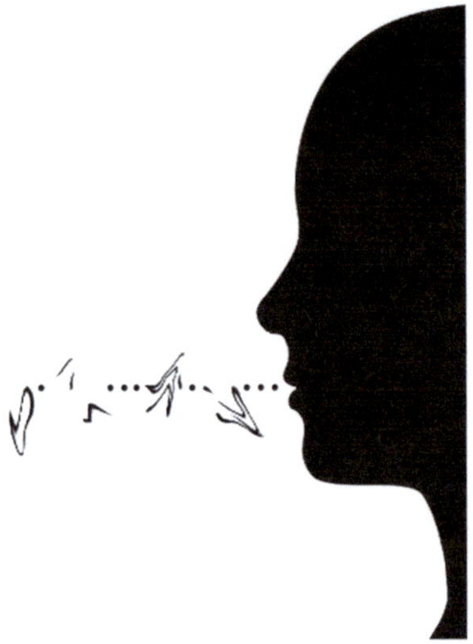

Notes:

- Occasionally speech arrest will not be elicited using a TMS intensity of 100% RMT. If results are not apparent, increase stimulus intensity in 5% increments until speech arrest occurs consistently. For safety, it is not recommended to exceed 150% RMT (*Sect. 2.2*); however, 120% RMT may be more conservative. If the participant reports discomfort at 100% RMT, decrease to 80–90% RMT.

4.5.2 Step for Inducing Covert Speech Arrest

Step 1: Attach EMG electrodes to the right FDI using a belly-tendon montage with a ground on the bony wrist (*Sect. 3.1*).

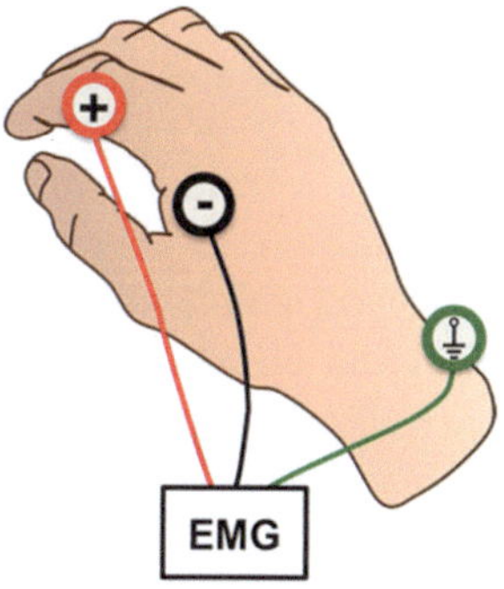

Step 2: Determine the FDI motor hotspot (*Sect. 3.3*). Remember that when determining the hotspot, one should start at a low intensity (i.e., below threshold) and increase until responses are large and consistently observed (i.e., above threshold). The hands should be completely at rest (i.e., no EMG activity) for the hotspot.

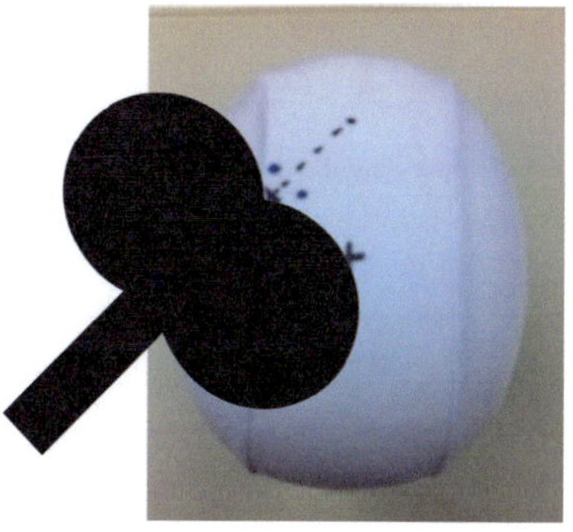

Step 3: Determine the RMT for the FDI hotspot (*Sect. 3.4*). For this example, we will assume an RMT of 50% MSO.

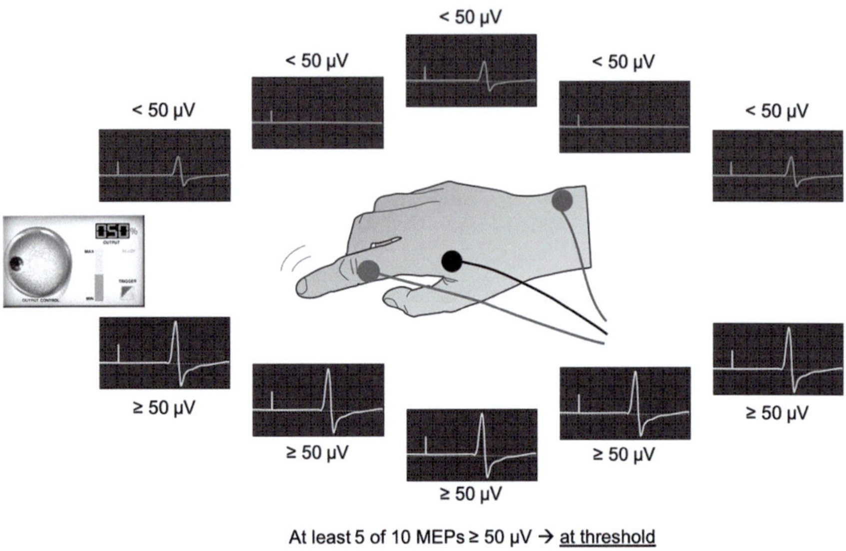

Step 4: Set TMS intensity to 100% RMT. For this example, this would equal 50% MSO.

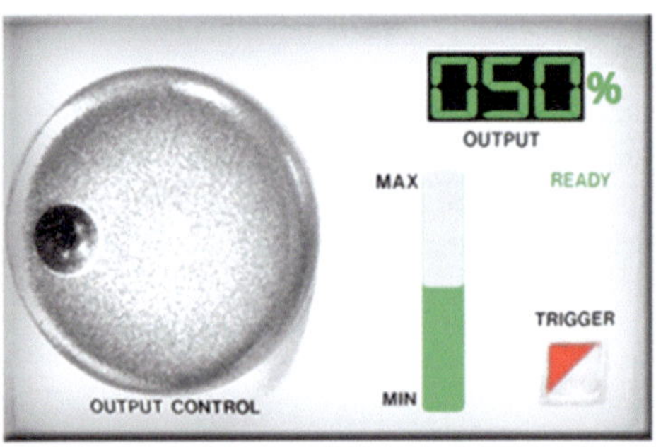

Step 5: Have the participant sit comfortably with a computer monitor positioned ~50 cm from their eyes. Give the subject a response box to collect responses.

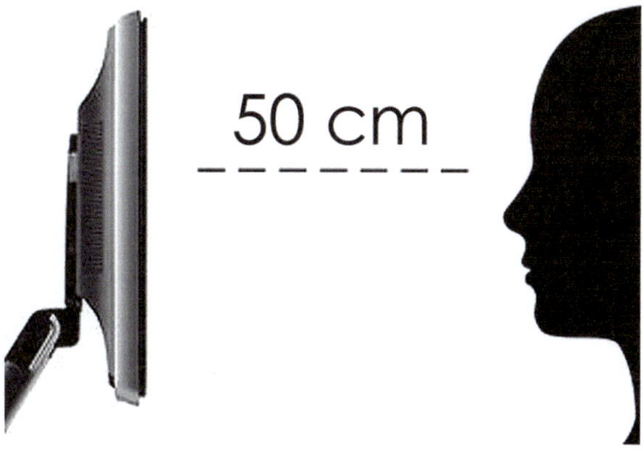

Step 6: Using any presentation software, display a word that contains three, four, or five syllables. Ask the participant to determine the number of syllables without speaking (i.e., covert speech) and to indicate their decision as quickly as possible by pressing the corresponding response button.

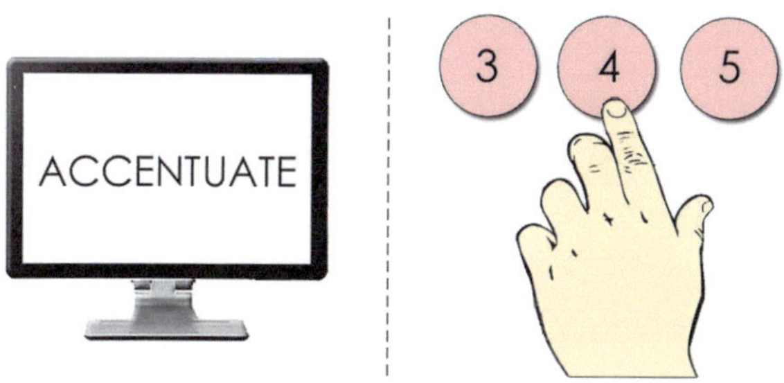

Step 7: Repeat Step 6 with 20 different words (ranging in syllable count) and record the response times for each trial. This represents the no-TMS baseline/control condition.

Step 8: From the right FDI hotspot, measure approximately 5 cm anterior and 5 cm lateral (toward the cheekbone). This site falls around EEG coordinate F7 and roughly corresponds to the middle portion of the temple. Place the center of the coil on this site with the handle pointing toward the back of the head to minimize activation of sensitive facial muscles.

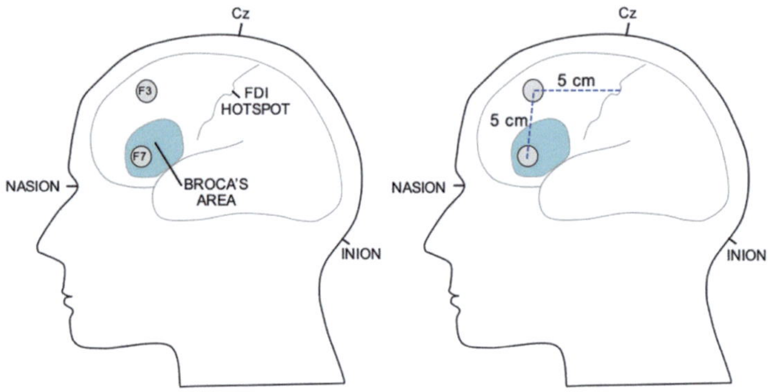

Step 9: Apply a 5 Hz rTMS train (200 ms inter-pulse interval) for 3 s (15 pulses total). During the rTMS train, have the subject repeat the syllable counting task (Steps 6 and 7).

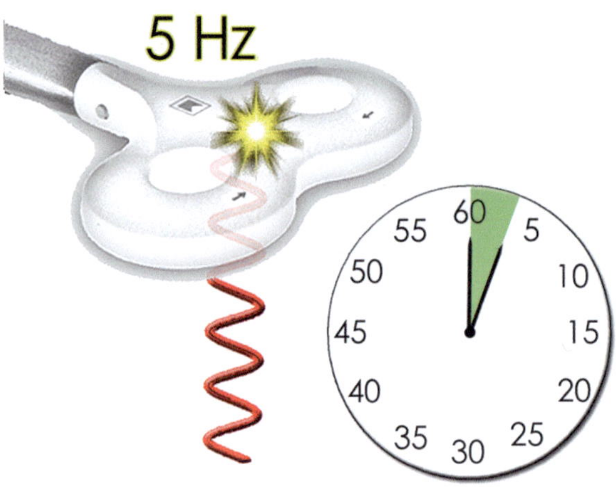

Step 10: Repeat Step 9 with 20 different words (ranging in syllable count) and record the response times for each trial. This represents the active TMS condition.

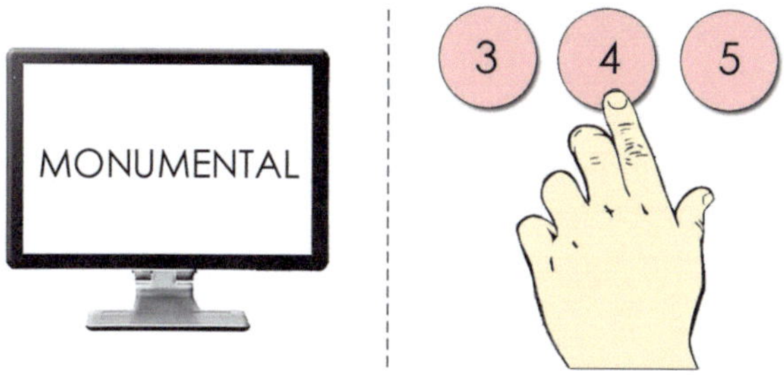

Step 11: Average and compare responses for the no-TMS and TMS conditions. An increase in response times for the TMS condition compared to the no-TMS condition is an indication of a covert (internal) speech arrest.

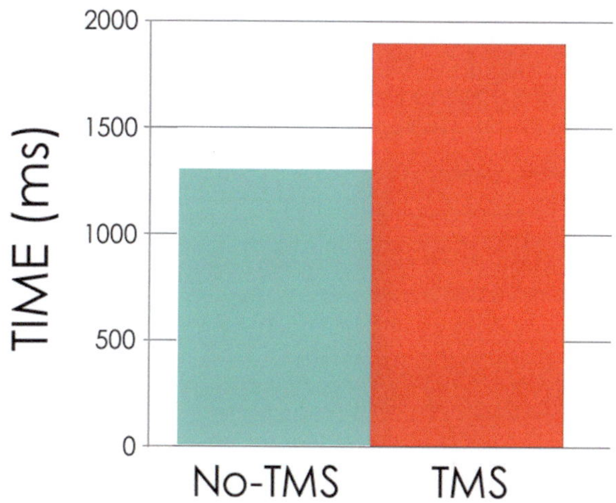

Notes:

- As with overt speech arrest described above, occasionally speech arrest will not be elicited using a TMS intensity of 100% RMT. If results are not apparent, increase stimulus intensity in 5% increments until speech arrest occurs consistently. For safety, it is not recommended to exceed 150% RMT (*Sect. 2.2*); however, 120% RMT may be more conservative. If the participant reports discomfort at 100% RMT, decrease to 80–90% RMT.

- Mapping of brain areas where covert speech arrest occurs can be performed similar to overt speech arrest and motor mapping (*Sect. 4.4*). To map the area of speech arrest, create a 5 × 5 cm grid centered on the site in Step 8, marking points at 1 cm intervals (alternatively, a 3 × 3 cm grid in 0.5 cm increments can be used). If using a neuronavigation system to perform the mapping, mark these locations on the subject's MRI. Repeat Steps 9 and 10 for each site. Average and compare response time for each site and determine map dimensions accordingly.

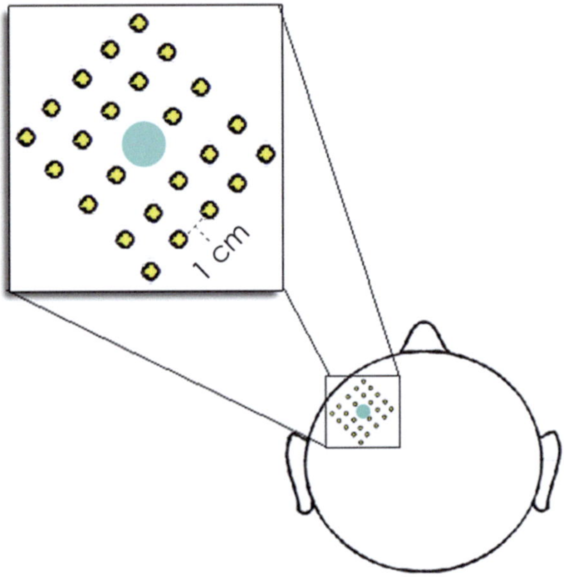

Further Reading

Aziz-Zadeh, L., Cattaneo, L., Rochat, M., & Rizzolatti, G. (2005). Covert speech arrest induced by rTMS over both motor and nonmotor left hemisphere frontal sites. Journal of Cognitive Neuroscience, 17(6), 928-938.

Carroll, T. J., Riek, S., & Carson, R. G. (2001). Reliability of the input–output properties of the cortico-spinal pathway obtained from transcranial magnetic and electrical stimulation. Journal of Neuroscience Methods, 112(2), 193-202.

Chen, R., Lozano, A. M. & Ashby, P. Mechanism of the silent period following transcranial magnetic stimulation. Evidence from epidural recordings. *Exp Brain Res* **128**, 539-542 (1999).

Daskalakis, Z. J. *et al.* An automated method to determine the transcranial magnetic stimulation-induced contralateral silent period. *Clin Neurophysiol* **114**, 938-944 (2003).

Fuhr, P., Agostino, R. & Hallett, M. Spinal motor neuron excitability during the silent period after cortical stimulation *Electroencephalography and Clinical Neurophysiology* **81**, 257-262 (1991).

Gangitano, M., Valero-Cabré, A., Tormos, J. M., Mottaghy, F. M., Romero, J. R., & Pascual-Leone, Á. (2002). Modulation of input–output curves by low and high frequency repetitive transcranial magnetic stimulation of the motor cortex. Clinical Neurophysiology, 113(8), 1249-1257.

Groppa S, Oliviero A, Eisen A, Quartarone A, Cohen LG, Mall V, et al. A practical guide to diagnostic transcranial magnetic stimulation: report of an IFCN committee. Clin Neurophysiol. 2012 May;123(5):858–82.

Hess, C.W. (2005). Central motor conduction time and its applications. In: M. Hallet and Chokroverty (Eds). *Magnetic Stimulation in Clinical Neurophysiology*. 2nd Ed. Elsevier.

Inghilleri, M., Berardelli, A., Cruccu, G. & Manfredi, M. Silent period evoked by transcranial stimulation of the human cortex and cervicomedullary junction. *J Physiol* **466**, 521-534 (1993).

Karhu, J., Hannula, H., Laine, J., & Ruohonen, J.. (2014). Navigated TMS – principles and protocol for mapping motor cortex. In Rotenberg, A., Horvath, J.C., & Pascual-Leone, A. (Eds.) *Neuromethods: Transcranial magnetic Stimulation*. New York: Springer Press.

Meyer, B. U., Roricht, S., Grafin von Einsiedel, H., Kruggel, F. & Weindl, A. Inhibitory and excitatory interhemispheric transfers between motor cortical areas in normal humans and patients with abnormalities of the corpus callosum. *Brain* **118 (Pt 2)**, 429-440 (1995).

Michelucci, R., Valzania, F., Passarelli, D., Santangelo, M., Rizzi, R., Buzzi, A. M., ... & Tassinari, C. A. (1994). Rapid-rate transcranial magnetic stimulation and hemispheric language dominance usefulness and safety in epilepsy. Neurology, 44(9), 1697-1697.

Pascual-Leone, A., Gates, J. R., & Dhuna, A. (1991). Induction of speech arrest and counting errors with rapid-rate transcranial magnetic stimulation. Neurology, 41(5), 697-702.

Pitcher, J. B., Ogston, K. M., & Miles, T. S. (2003). Age and sex differences in human motor cortex input–output characteristics. The Journal of Physiology, 546(2), 605-613.

Rossini PM, Burke D, Chen R, Cohen LG, Daskalakis Z, Di Iorio R, et al. Non-invasive electrical and magnetic stimulation of the brain, spinal cord, roots and peripheral nerves: Basic principles and procedures for routine clinical and research application. An updated report from an I.F.C.N. Committee. Clin Neurophysiol. 2015 Jun;126(6):1071–107.

Tarapore, P.E. (2014). Speech mapping with transcranial magnetic stimulation. In Rotenberg, A., Horvath, J.C., & Pascual-Leone, A. (Eds.) *Neuromethods: Transcranial magnetic Stimulation*. New York: Springer Press.

Udupa, K., Chen, R., 2013. Brain Stimulation: Chapter 31. Central motor conduction time. Elsevier.

Wassermann, E. M., Fuhr, P., Cohen, L. G. & Hallett, M. Effects of transcranial magnetic stimulation on ipsilateral muscles. *Neurology* **41**, 1795-1799 (1991).

Wilson, S. A., Thickbroom, G. W. & Mastaglia, F. L. An investigation of the late excitatory potential in the hand following magnetic stimulation of the motor cortex. *Electroencephalography and Clinical Neurophysiology* **97**, 55-62 (1995).

Chapter 5
Advanced Neurophysiology Measures

Abstract This chapter describes paired- and triple-pulse techniques aimed at probing corticospinal physiology. Most commonly, TMS applications have used a pair of sequential pulses in the same brain location, or in two different locations. For paired-pulse TMS using a single coil targeting one brain region, pulses are delivered in close succession, requiring the stimulator to have greater than a single capacitor and precise timing control. By observing MEP amplitude change because of a prior conditioning pulse, local intracortical inhibitory and facilitatory influence on the corticospinal system can be examined. A similar approach can be applied using two coils, each in separate brain locations, to examine the influence of one brain region over the other. Single-pulse TMS can also be paired with peripheral nerve stimulation, wherei. TMS is delivered to unilateral primary motor cortex, and temporally associated with electrical stimuli in the periphery. The techniques described in this section have a range of applications including to investigate the pathophysiology of neurological or psychiatric disease, or the effect of intervention.

D. J. Edwards et al., *A Practical Manual for Transcranial Magnetic Stimulation*, https://doi.org/10.1007/978-3-031-62304-2_5

5.1 Paired-Pulse TMS to One Brain Region

In paired-pulse TMS paradigms, two TMS pulses are applied at a given interstimulus interval (ISI) to measure the effect of the first pulse, termed the conditioning stimulus (CS), on the second pulse, termed the test stimulus (TS). This "conditioning effect" is compared to the response obtained by applying the unconditioned TS (i.e., a single TMS pulse). This section will discuss paired-pulse TMS to a single brain region using a single coil to deliver both the CS and TS. This approach can be used to probe the excitability of local intracortical circuits and inform on the balance of GABAergic-mediated inhibition and glutamatergic-mediated facilitation. A parallel approach involved applying the CS and TS to two different brain regions using two different coils and will be discussed in Sect. 5.2.

Differential paired-pulse effects can be observed by changing the ISI and the intensity of the CS. The most common paired-pulse TMS protocols are short-interval intracortical inhibition (SICI), intracortical facilitation (ICF), and long-interval intracortical inhibition (LICI). The figure below lists the parameters and expected response of these three protocols.

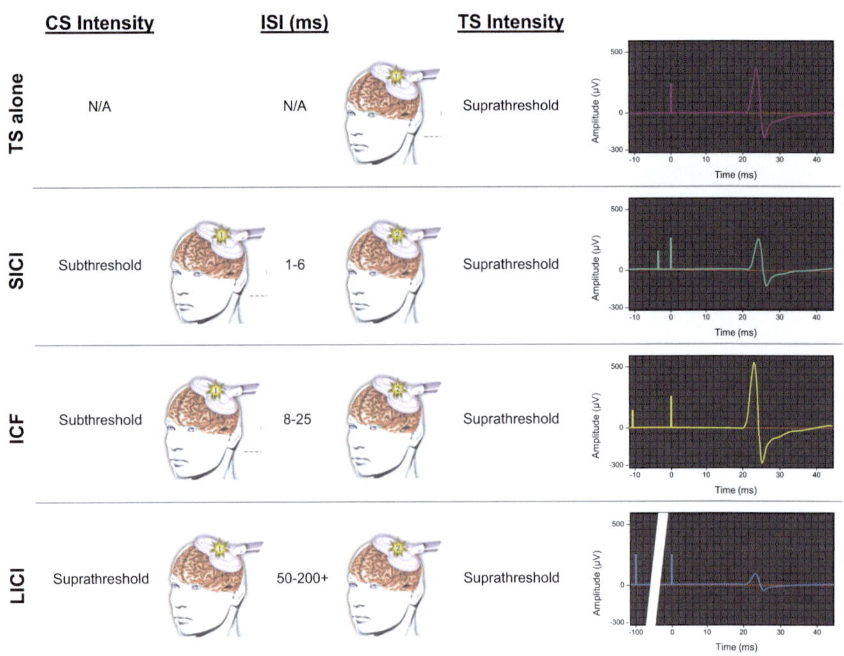

As noted in the figure above, paired-pulse conditioning effects can be observed using a range of ISIs for each protocol. Under typical conditions, short ISIs (1–6 ms) generate SICI, medium ISIs (8–25 ms) generate ICF, and long ISIs (50–200+ ms) generate LICI. One approach to administering paired-pulse TMS protocols uses a fixed ISI for each protocol, while the other uses a range of ISIs to create a paired-pulse curve. This section will focus on the fixed-ISI approach and briefly discuss the paired-pulse curve.

When applied to M1 and using EMG to measure the conditioned response, SICI and LICI manifest as a reduction in MEP amplitude, while ICF manifests as an increase in MEP amplitude. Pharmacologic studies suggest $GABA_A$ agonists and glutamatergic antagonists enhance inhibition and decrease facilitation, respectively, for SICI and ICF, while ion-channel blockers show no effect. Drugs that act on $GABA_B$ receptors show the strongest effects on LICI. Accordingly, SICI, LICI, and ICF may provide useful information on GABAergic and glutamatergic neurotransmission in the cortex. Studies in patients have demonstrated that varied neurological and psychiatric disorders (e.g., epilepsy, Alzheimer's disease, Parkinson's disease, amyotrophic lateral sclerosis) can influence the inhibitory and facilitatory effects of paired-pulse TMS. When combined with the pharmacologic data, these findings have been used to inform on alterations in GABAergic and glutamatergic neurotransmission in these and other neuropsychiatric conditions. When comparing the conditioning effects of paired-pulse TMS protocols between groups (e.g., Alzheimer's versus healthy), the use of paired-pulse TMS curves (as opposed to set ISIs) can be useful in interpreting group differences.

The protocols for SICI, ICF, and LICI all deliver the TS at a suprathreshold intensity to activate layer-V pyramidal neurons that give rise to the corticospinal tract. For LICI, the CS is also applied at a suprathreshold intensity (often at the same intensity as the TS) and will elicit an MEP. This MEP can be used as the unconditioned response (to compare against the conditioned response) when the CS and TS of LICI are applied at the same intensity. By contrast, the CS for SICI and ICF is always applied at a subthreshold intensity. This means that the CS of SICI or ICF will not be sufficient to elicit an MEP, though it will still modulate intracortical circuits that converge on layer-V pyramidal neurons. Consequently, the conditioned responses for SICI and ICF must be compared to a separate unconditioned MEP response. The conditioning response of SICI, ICF, and LICI can all be expressed in terms of the % change from the unconditioned TS only using the formula below:

$$\frac{\left(\textbf{conditioned TS} - \textbf{unconditioned TS}\right)^{*} \textbf{100}}{\left(\textbf{unconditioned TS}\right)}$$

Using this formula, negative values represent inhibition and positive values represent facilitation, while a value of 0 represents no change.

As an alternative, the conditioned TS can be expressed as a ratio of the unconditioned TS using the formula below:

$$\frac{\textbf{conditioned TS}}{\textbf{unconditioned TS}}$$

Using this formula, values below 1 represent inhibition and values above 1 represent facilitation. A value of 1 represents no change.

Below, we will demonstrate paired pulse using the protocol for SICI and discuss the modification necessary to perform ICF and LICI.

Note As the dual pulses are generated in rapid succession through a single coil, this procedure will require a TMS machine with multiple charging units and a pulse timing mechanism.

5.1.1 Steps for Measuring Intracortical Inhibition and Facilitation in Left M1 with the Right FDI as the Target Muscle

Step 1: Attach EMG electrodes to the right FDI using a belly-tendon montage with a ground on the bony wrist (*Sect. 3.1*).

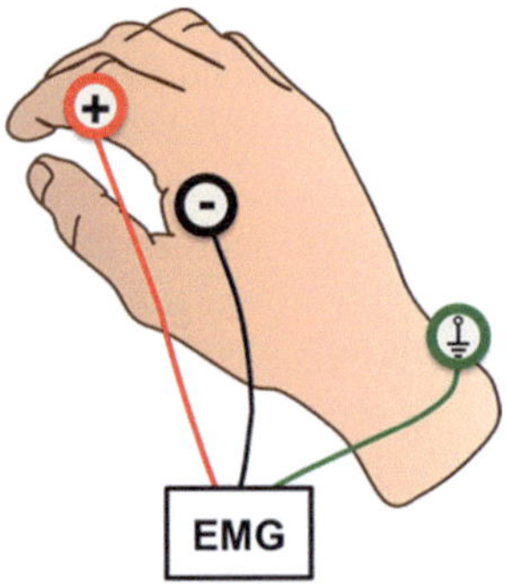

Step 2: Determine the FDI motor hotspot (*Sect. 3.3*). Remember that when determining the hotspot, one should start at a low intensity (i.e., below threshold) and increase until responses are large and consistently observed (i.e., above threshold). The hands should be completely at rest (i.e., no EMG activity) for the hotspot.

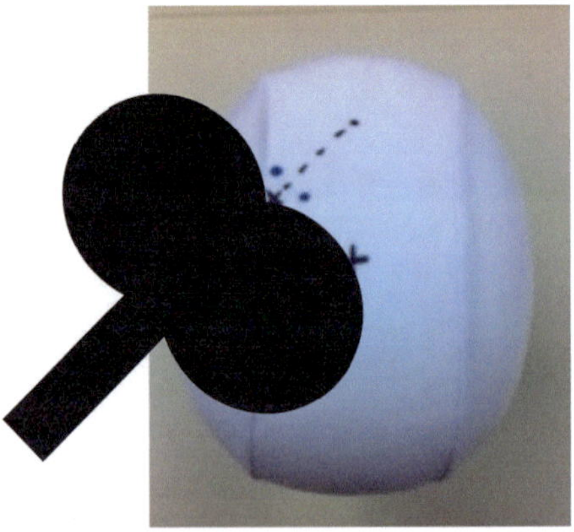

Step 3: Determine the RMT for the FDI hotspot (*Sect. 3.4*). For this example, we will assume an RMT of 50% MSO.

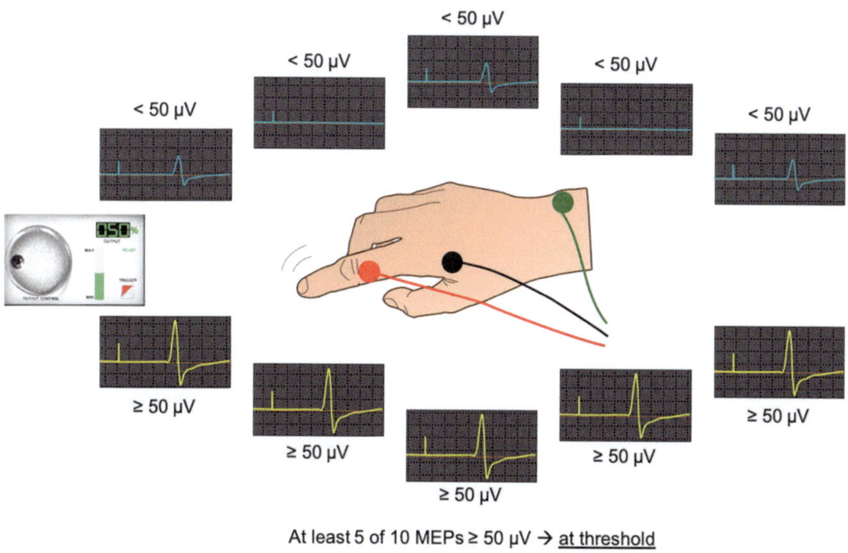

Step 4: With intensity set to 120% RMT (60% MSO in this example), deliver 10 single pulses (spaced every 6–10 s). The average MEP amplitude will serve as the unconditioned (TS alone) response.

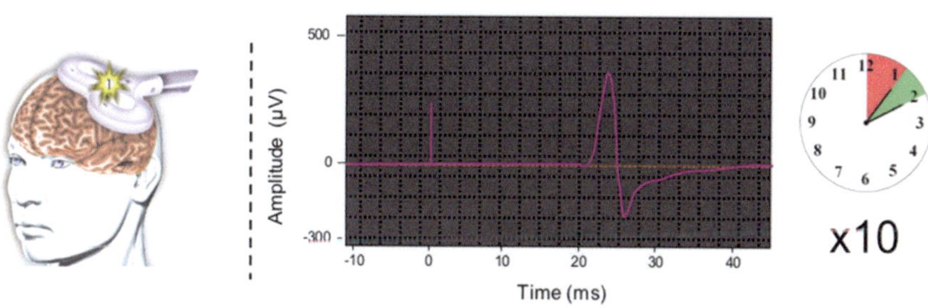

Step 5: With TS at 120% RMT as above, set the CS to 80% RMT. For this example, the TS is 60% MSO, and the CS is 40% MSO.

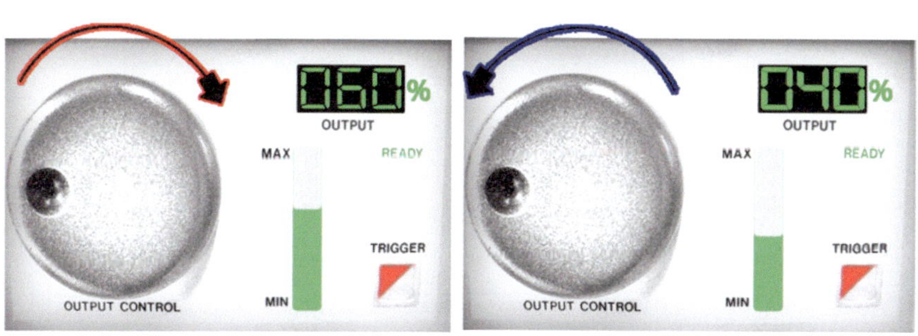

Step 6: To assess SICI, choose an ISI from 1 to 6 ms. For this example, we will use a 3-ms ISI.

Step 7: Deliver 10 trials (spaced every 6–10 s). Average the trials and express them as the % change (or ratio) of the unconditioned MEPs from Step 4. This will be the SICI response.

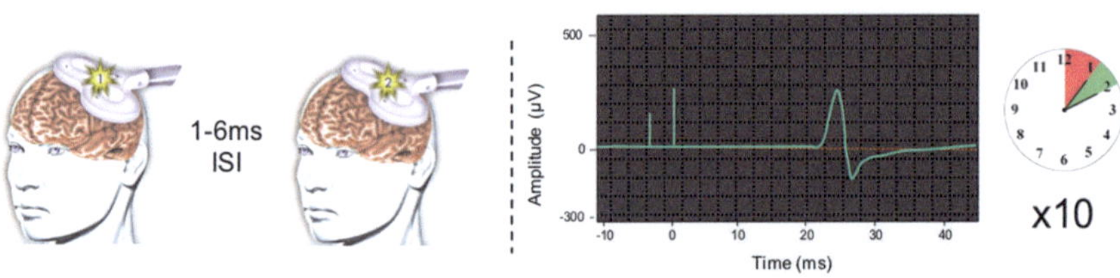

Step 8: To assess ICF, follow Steps 1–5 and choose an ISI between 8 and 25 ms. For this example, we will use a 12-ms ISI.

Step 9: Deliver 10 trials (spaced every 6–10 s). Average the trials and express them as the % change (or ratio) of the unconditioned MEPs from Step 4. This will be the ICF response.

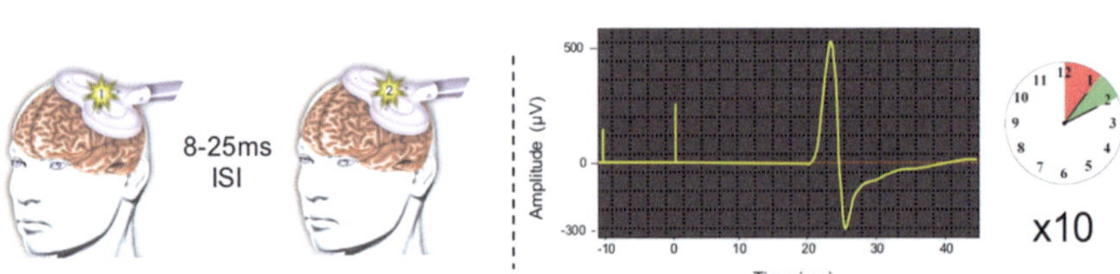

Step 10: To assess LICI, follow Steps 1–4. Set the CS and the TS to 120% RMT (60% MSO in this example).

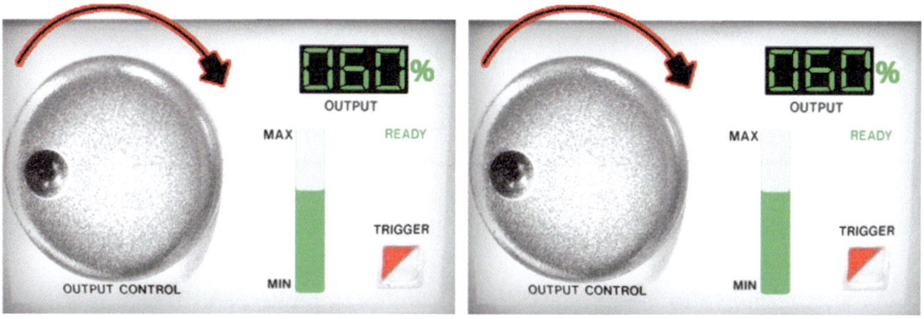

Step 11: Choose an ISI between 50 and 200 ms. For this example, we will use a 100-ms ISI.

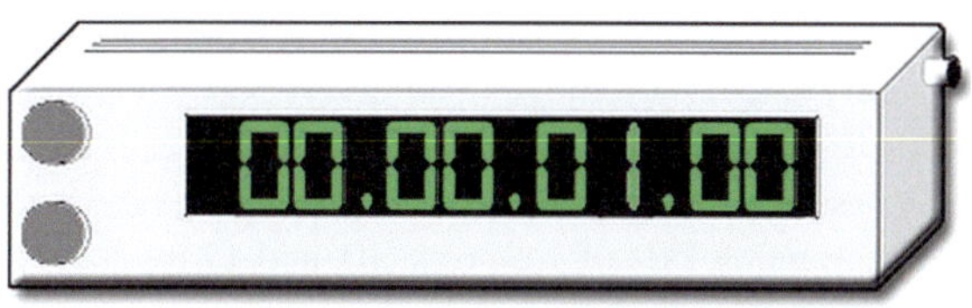

Step 12: Deliver 10 trials (spaced every 6–10 s). Average the trials and express them as the % change (or ratio) of the unconditioned MEPs from Step 4 (or the average MEP from the CS). This will be the LICI response.

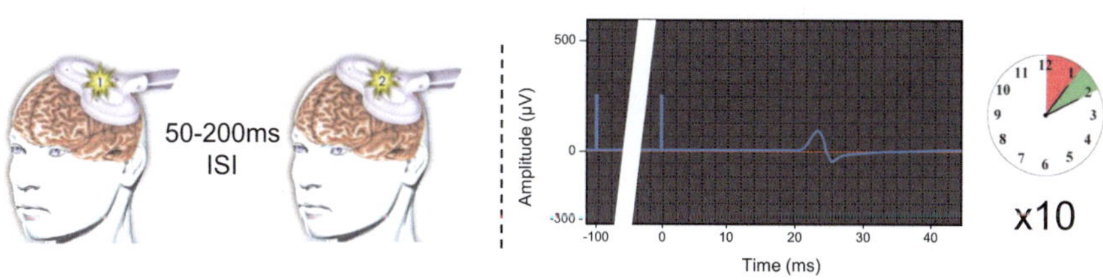

Notes:

- These protocols can theoretically be measured for any cortically represented target muscle.
- While it is theoretically possible to perform paired-pulse TMS during a low-level muscle contraction with intensities based on AMT (*Sect. 3.5*), the mechanisms are less well understood, and the effects may not be comparable to those obtained with the target muscle at rest.
- An alternative to setting the TS intensity based on RMT, intensity is adjusted to elicit a specific MEP amplitude value (e.g., 1 mV). This practice can be used to standardize the TS when performing paired-pulse protocols under conditions (e.g., diseases or pharmacologic interventions) that may change excitability. This will also help avoid floor and/or ceiling effects potentially generated by pharma-

ceutical or other interventions. However, this process requires more precision and time that should be accounted for when using this option. Further, conditions that result in low excitability may make it difficult to obtain a 1-mV MEP in all participants. One option is to use an I/O curve (*Sect. 4.1*) to estimate the intended MEP amplitude (e.g., using the intensity that corresponds to the 50% maximal amplitude of the curve).

- MEP amplitude can spontaneously drift with time (due to any number of biological and environmental factors). Accordingly, when possible, single- (TS alone) and paired-pulse measures should be interleaved (rather than perform each protocol as a separate block). This will help diminish any differences caused by natural excitability drift and or changes in coil position.
- As an alternative to using one ISI per protocol, a range of ISIs can be used to generate a paired-pulse curve. This is a form of I/O curve (*Sect. 4.1*) that evaluates the conditioning effects (y-axis) of a range of ISIs (x-axis). While a curve can be generated for any paired-pulse protocol, it is more common for SICI and ICF, which share the same CS and TS intensities and differ only in their ISIs. Accordingly, it is possible to demonstrate the shift from inhibition to facilitation in a single curve by selecting a range of intervals between 1 and 6 ms for SICI and 8–25 ms for ICF. When MEP amplitude is plotted as a function of ISI, the results typically resemble a sigmoidal curve. The figure below shows the hypothetical effects from using ISIs of 1, 3, 6, 8, 10, and 12 ms.

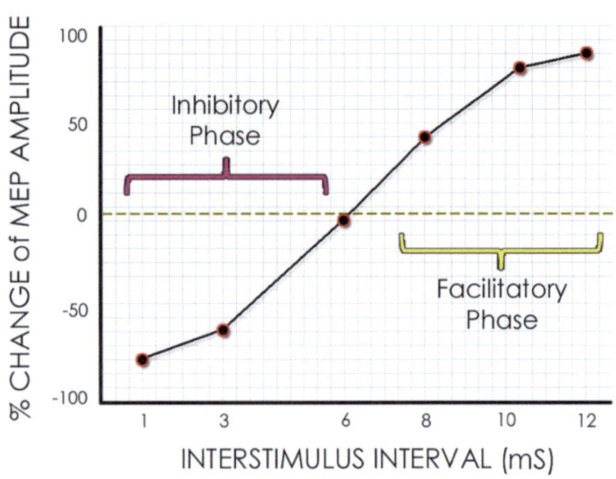

- This approach provides a more complete picture of the magnitude and timing of GABAergic and glutamatergic mechanisms that underlie intracortical inhibition and facilitation and their interactions. Patient studies have demonstrated that varied neurological and psychiatric disorders can influence the inhibitory and/or facilitatory phases of a paired-pulse curve.
- When applying a paired-pulse curve, it is advisable to randomize the order in which the ISIs are applied to avoid any possible cumulative effects of a stepwise ISI increase/decrease. Additional ISIs can be used to increase the specificity of the curve.

- Although the parameters outlined above are typical of paired-pulse studies, there are additional intensities and/or ISIs to investigate intracortical inhibitory and facilitatory interactions (see the figure at the beginning of this section).
- Conditioning pulse intensity can change the effect on test-pulse MEP amplitude at a given ISI. At inhibitory ISIs, increased condition pulse intensity toward or above RMT may cause facilitation.

5.2 Paired-Pulse TMS to Two Brain Regions

The general paired-pulse TMS paradigm that utilizes a single coil to target a single brain region (*Sect. 5.1*) can be adapted to evaluate the interactions between two brain structures. This is achieved by applying the CS and TS with different coils to two spatially separate brain regions, which may be different regions with the same cerebral hemisphere (intrahemispheric paired-pulse TMS) or homologous regions in left and right cerebral hemispheres (interhemispheric paired-pulse TMS). It is also possible to apply this approach to the cerebellum and cerebral cortex. In this example (cerebellar-brain inhibition; CBI), the excitability of the dentato-thalamo-cortical pathway can be assessed by evaluating the interactions between a CS to the posterior lobe of the cerebellum and a TS to the contralateral M1, likely via the ventrolateral nucleus of the thalamus. The figure below illustrates the most common dual-coil protocols, including interhemispheric inhibition (IHI), intrahemispheric (premotor to M1) facilitation, and CBI.

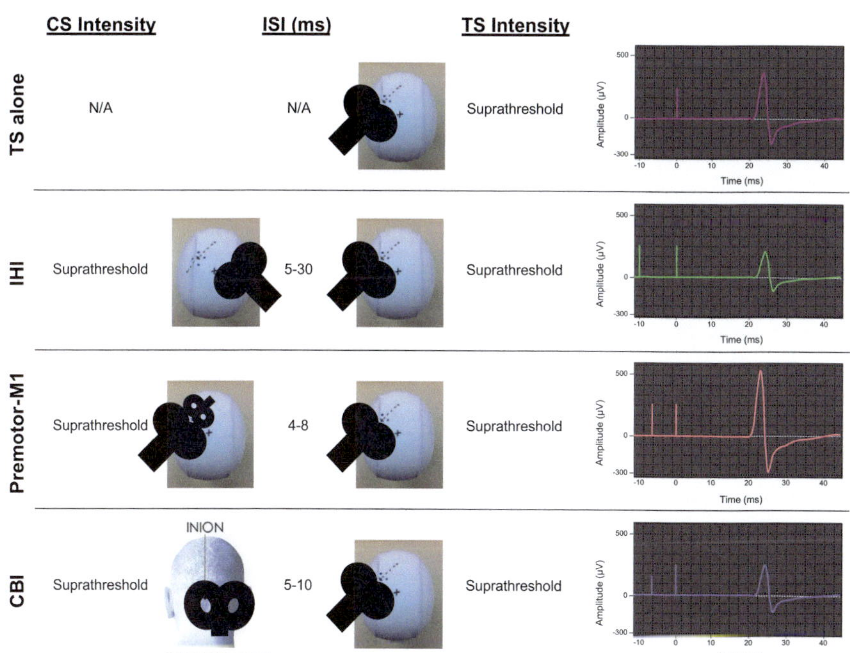

This section will focus on IHI as it is the most well-understood dual-coil paired-pulse paradigm. The mechanisms of IHI are rooted in the phenomenon that for many regions of the cortex, including M1, pyramidal cells in layers II and III send axons to the homologous region via the corpus callosum. These callosal projections are excitatory but they synapse on local inhibitory interneurons leading to a net inhibition on the contralateral region. Inhibitory balance between hemispheres is an important feature for many brain regions. In M1, it serves to inhibit mirror movements between left and right muscles and also plays an important role in motor stroke deficits and recovery. We can measure this inhibitory effect, i.e., IHI, by applying a CS to one M1 and a TS to the contralateral M1 and observing the change in MEP compared to a TS alone.

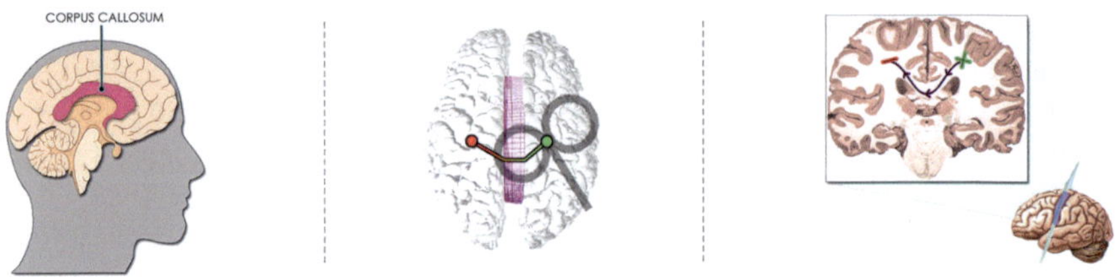

As with paired-pulse TMS to one brain region (*Sect. 5.1*), the CS and TS in these dual-coil techniques are delivered in close sequential order. The conditioning effects are thought to be mediated through neural circuits connecting the two regions, which may involve GABAergic inhibition and/or glutamatergic facilitation. As measured via EMG, inhibition will manifest as diminished MEP values and facilitation as enhanced MEP values. The conditioning response can be expressed as either (1) the % change from the unconditioned TS or (2) a ratio of the unconditioned TS:

$$(1)\ \frac{(\textbf{conditioned TS} - \textbf{unconditioned TS})^{*}\ \textbf{100}}{(\textbf{unconditioned TS})} \qquad (2)\ \frac{\textbf{conditioned TS}}{\textbf{unconditioned TS}}$$

As with other paired-pulse interactions, the conditioning effects can be observed under a range of intensities and ISIs. Depending on the nature of the connectivity between the targeted brain structures, these parameters will yield different conditioning effects. For example, when targeting M1 bilaterally, a longer ISI (5–30 ms) will often generate inhibition, i.e., IHI. It is possible to observe facilitation between left and right M1 using a short ISI (3–5 ms).

For intrahemispheric and cerebellar to brain interactions, the mechanisms are less well understood. When evaluating the effects of a CS to the premotor cortex on a TS to M1, it appears that the intensity of the CS plays a larger role in determining the conditioning effects. For example, at a short ISI (4–8 ms), a subthreshold CS will often generate inhibition, while suprathreshold CS will often generate facilitation. For cerebellar–brain interactions, a short ISI (5–10 ms) with a suprathreshold intensity for both CS and TS will often generate inhibition.

Step-by-step instructions to assess IHI are shown below. At the end, we will briefly describe how to adjust the protocol to probe intrahemispheric (premotor to M1) interactions and CBI as well as how to apply any dual-coil protocol as a paired-pulse curve.

5.2.1 *Steps for Paired-Pulse Across Two Brain Regions: Interhemispheric Inhibition of Left M1 by Right M1 with the Right FDI as the Target Muscle*

Step 1: Attach EMG electrodes to the left and right FDI using a belly-tendon montage with a ground on the bony wrist (*Sect. 3.1*).

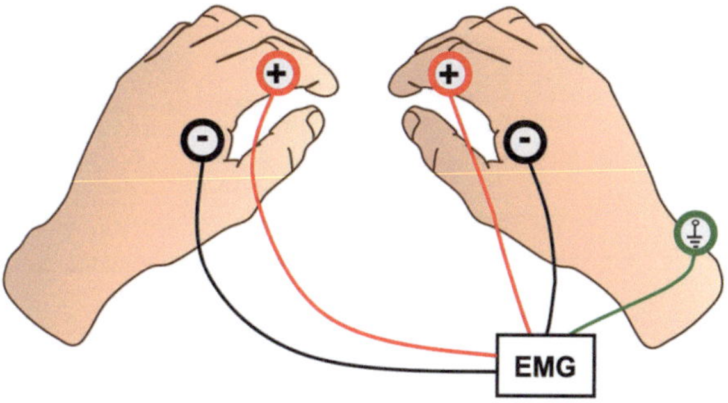

Step 2: Determine the FDI motor hotspot (*Sect. 3.3*) for both hemispheres. Remember that when determining the hotspot, one should start at a low intensity (i.e., below threshold) and increase until responses are large and consistently observed (i.e., above threshold). The hands should be completely at rest (i.e., no EMG activity) for the hotspot.

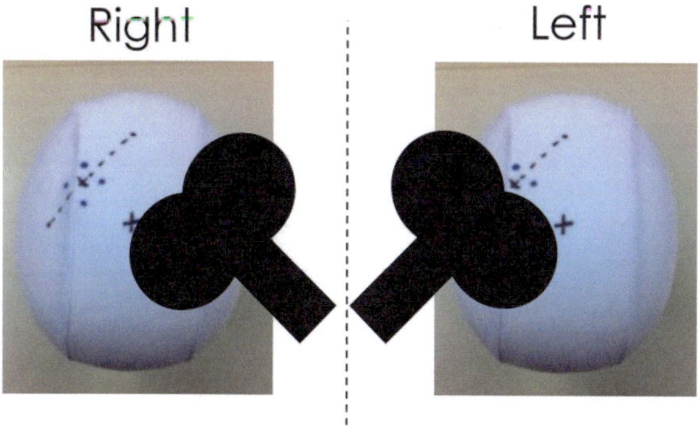

Step 3: Determine the RMT for the FDI hotspot (*Sect. 3.4*). Do this for both left and right hemispheres. For this example, we will assume the RMT equals 50% MSO in both hemispheres. In reality, left and right RMT may differ.

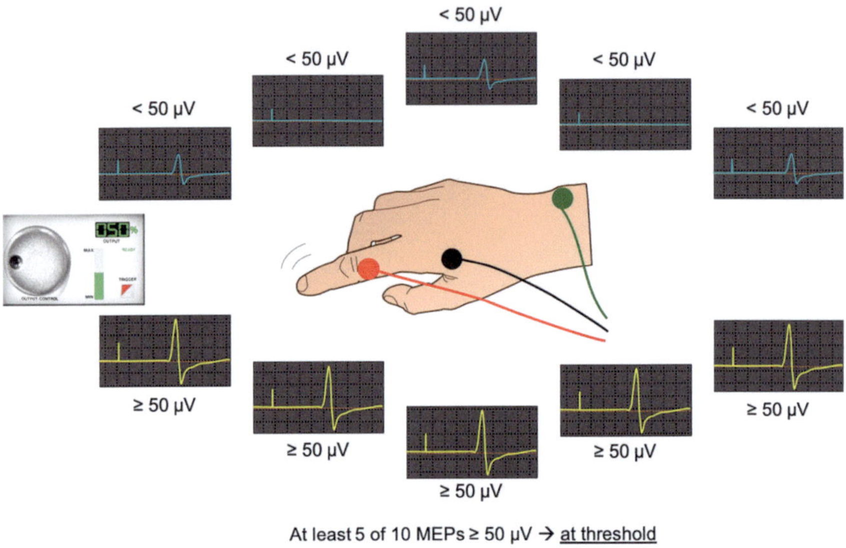

Step 4: With intensity set to 120% RMT (60% MSO in this example), deliver 10 single pulses (spaced every 6–10 s) to the left M1 hotspot. The average MEPs will serve as the unconditioned (TS alone) response.

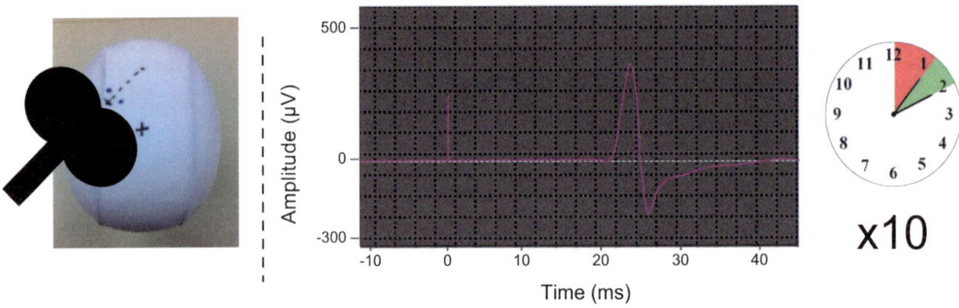

Step 5: Set the CS and TS intensity to 120% RMT. Since the intensity is the same, the order of the pulses determines the direction of the conditioning effects. In this example, we are applying the TS to the left hemisphere and measuring MEPs from the right FDI. This will test the conditioning effects of the right hemisphere on the left hemisphere. To change the direction of effect, reverse the order of CS and TS.

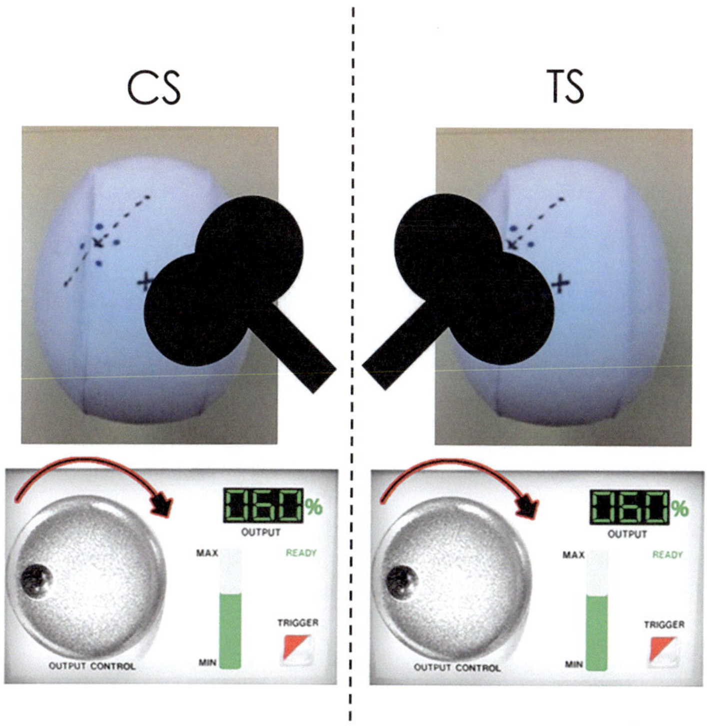

Step 6: Choose an ISI between 6 and 30 ms. For this example, we will use a 10 ms ISI.

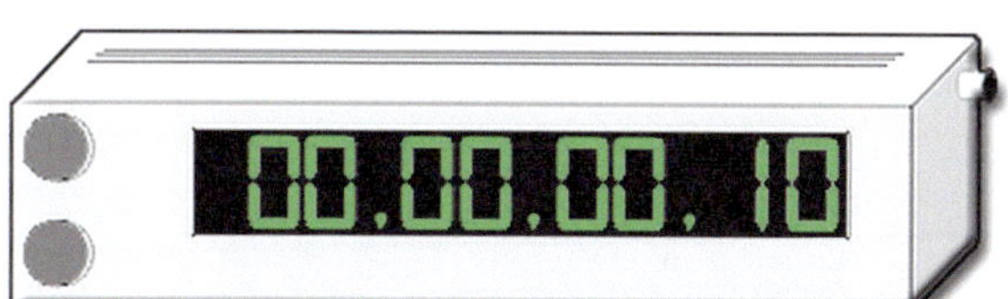

Step 7: Deliver 10 trials (spaced every 6–10 s). Average the trials and express them as the % change (or ratio) of the unconditioned MEPs from Step 4. This will be the IHI response.

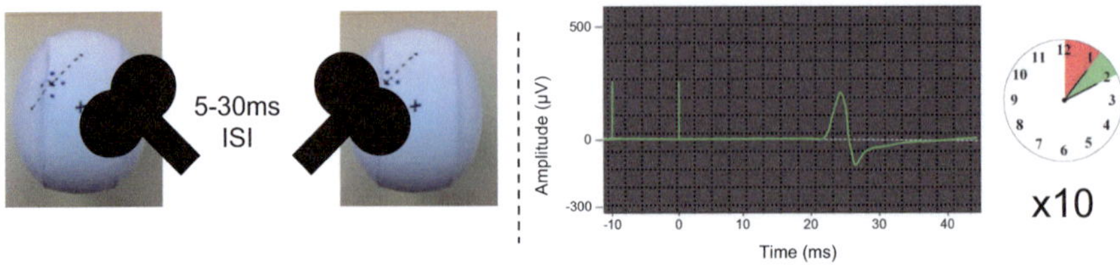

Notes:

- As the dual pulses are generated in rapid succession through two different coils, these procedures will require a TMS machine with multiple charging units, multiple coil outputs, and a pulse timing mechanism.
- An alternative approach to setting the TS intensity based on RMT is to set the intensity to elicit a specific MEP amplitude value (e.g., 1 mV). This practice can be used to standardize the TS when performing paired-pulse protocols under conditions (e.g., diseases or pharmacologic interventions) that may change excitability. This will also help avoid floor and/or ceiling effects that may result from pharmaceutical or other interventions. However, this process requires more precision and time that should be accounted for when utilizing this option. Furthermore, it may be difficult or impossible to obtain a 1-mV MEP in certain conditions that decrease cortical excitability or impair the integrity of the corticospinal tract. Due to these challenges, we recommend first using an I/O curve (*Sect. 4.1*) to estimate the intended MEP amplitude as it will give the most complete picture of the range of corticospinal responses.
- MEP amplitude can spontaneously drift over time due to any number of biological and environmental factors. Accordingly, when possible, single- (TS alone) and paired-pulse measures should be interleaved (rather than perform each protocol as a separate block). This will help avoid effects due to hysteresis and diminish "block effects" caused by natural excitability drift and or changes in coil position between conditions.

- This dual-coil paired-pulse protocol can be adapted to assess intrahemispheric (i.e., premotor to M1) and cerebellar to M1 interactions. Compared to IHI, these protocols are less well defined, and their mechanisms are still an area of active research. In addition, these protocols require the use of specialized coils (*Sect. 1.4*).

    To assess premotor to M1 interactions, the location for the premotor target is approximately 3–4 cm rostral to the M1 hotspot. Consequently, there is little room to place two full-size (i.e., 70 mm wing diameter) figure-8-coils over both spots. It may be easier to use one or more coils that are smaller and/or coils with a specialized design to achieve the correct placement.
    Stimulating the cerebellum poses a different challenge. Since it is believed that the effects of CBI are due in part to stimulation of the Purkinje cells and/or deep cerebellar nuclei, it may be easier to reach these deeper targets using an angled (double-cone) coil.

- As illustrated in (*Sect. 5.1*), a range of ISIs can be used to generate a paired-pulse curve for any of the dual-coil protocols discussed above. This is a form of an I/O curve (*Sect. 4.1*) that evaluates the conditioning effects (y-axis) of a range of ISIs (x-axis).
- The ISI is not the only variable that can be systematically explored in a paired-pulse curve. One can also apply these protocols using a range of intensities of the CS. For example, when using a fixed ISI of 6 ms to investigate in premotor to M1 interactions, a subthreshold CS has been shown to lead to the inhibition of the TS, while a suprathreshold CS leads to facilitation.
- It is also possible to construct a multi-dimensional curve by varying both the CS intensity and the ISI in a systematic fashion. While this approach requires a complex design, it would provide a more comprehensive characterization of the neurophysiologic interactions.
- When administering a paired-pulse curve, it is advisable to randomize the order of the x-axis steps (ISI and/or CS intensity). This is to avoid any possible hysteresis or cumulative effects of a stepwise increase/decrease.
- To achieve greater specificity, increase the number of x-axis steps (ISI and/or CS intensity) that are tested.

5.3 Combining Peripheral and Central Stimulation

The previous two sections have focused on the ability of a TMS pulse, either to M1 (*Sect. 5.1*) or a distally connected brain region (*Sect. 5.2*), to condition a subsequent MEP response from TMS to M1. In this section, we will explore how TMS pulses to M1 can also be conditioned by electrical stimuli delivered to the peripheral nervous system (PNS). This is a means to test interactions between the central nervous system (CNS) and PNS. An electrical stimulus delivered to a peripheral nerve elicits action potentials that propagate along the nerve. There are two common protocols—short afferent inhibition (SAI) and triple-pulse stimulation technique (TST)—that combine a TS to M1 using TMS with one or more CS to the PNS using electrical stimuli delivered to mixed sensory/motor fibers (e.g., the median nerve).

5.3.1 *Short Afferent Inhibition*

SAI tests the interactions between sensory (afferent) and motor (efferent) systems. Similar to other paired-pulse protocols, a TS (a suprathreshold TMS pulse to M1) is preceded by a CS (an electrical stimulus delivered to a peripheral nerve). The timing of SAI (the ISI between the electrical CS and the electromagnetic TS) reflects the conduction time of the peripheral and central pathways. The sensory afferent can be recorded as an N20 somatosensory potential using EEG (a negative deflection of the EEG around 20 ms after an electrical stimulus to the median nerve of the wrist, reflecting the latency of afferent signals). Several studies have demonstrated that the conditioning effects of SAI occur ~1–4 ms after the N20 latency.

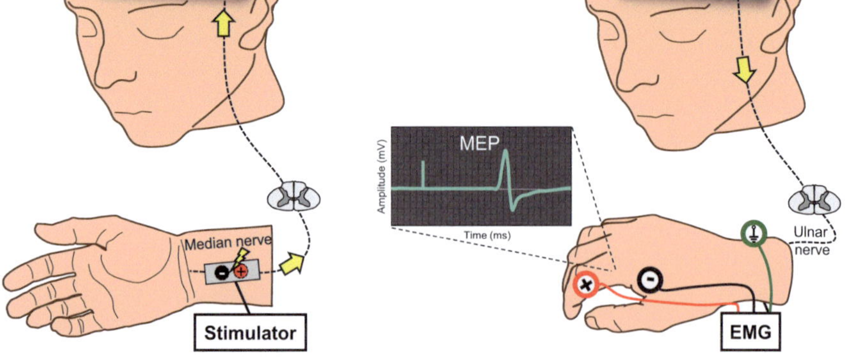

An electrical stimulus given to a mixed sensory/motor nerve, e.g., median nerve, elicits two types of afferent signals: an *orthodromic* signal along the sensory component, which will eventually reach the primary somatosensory cortex and an *antidromic* signal along the motor component, which will reach the spinal cord and may travel to M1. While the precise contributions of sensory and motor afferents in SAI remain to be elucidated, the net result is an inhibitory conditioning of M1. This inhibition can be measured as a reduction in MEP amplitude from a TS to M1. Studies using subdural recording electrodes over the upper cervical spinal cord suggest that SAI primarily involves a conditioning of I3 waves (*Sect. 1.5*), which are believed to reflect polysynaptic cortico-cortical connections.

Pharmacologic studies have demonstrated that the effects of SAI can be modulated by drugs that affect muscarinic receptors, suggesting that SAI can be used to provide an index of central cholinergic function. In line with this evidence, SAI has been found to decline with age and be further reduced in Alzheimer's disease.

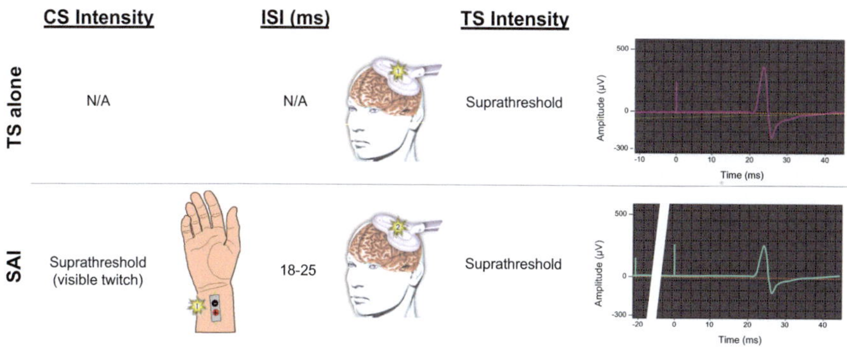

The conditioning response of SAI is expressed in terms of the % change from the unconditioned TS only using the formula below:

$$\frac{(\text{conditioned TS} - \text{unconditioned TS})^{*}\,100}{(\text{unconditioned TS})}$$

Using this formula, negative values represent inhibition and positive values represent facilitation, while a value of 0 represents no change.

As an alternative, you can express the conditioned TS as a ratio of the unconditioned TS using the formula below:

$$\frac{\text{conditioned TS}}{\text{unconditioned TS}}$$

Using this formula, values below 1 represent inhibition and values above 1 represent facilitation. A value of 1 represents no change.

Note This procedure will require a TMS machine, a peripheral electrical stimulator, and a pulse timing mechanism.

5.3.2 Steps for Assessing SAI in Left M1 Using Median Nerve Stimulation with the Right FDI as the Target Muscle

Step 1: Attach EMG electrodes to the right FDI using a belly-tendon montage with a ground on the bony wrist (*Sect. 3.1*).

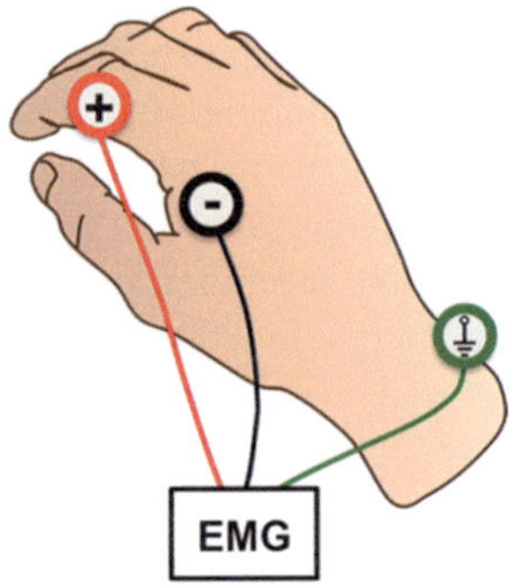

Step 2: Determine the FDI motor hotspot (*Sect. 3.3*). Remember that when determining the hotspot, one should start at a low intensity (i.e., below threshold) and increase until responses are large and consistently observed (i.e., above threshold). The hands should be completely at rest (i.e., no EMG activity) for the hotspot.

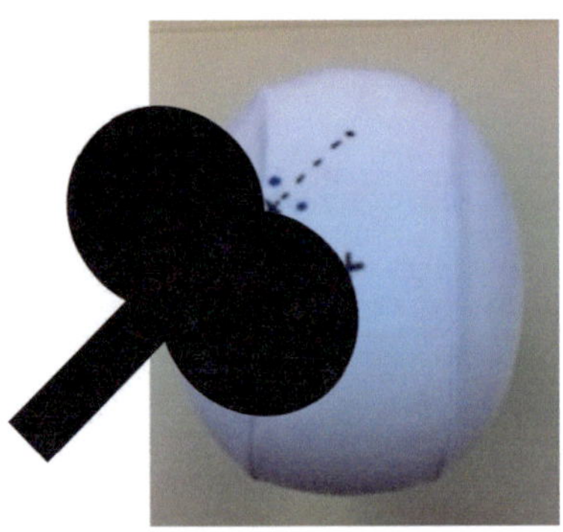

Step 3: Determine the RMT for the FDI hotspot (*Sect. 3.4*). For this example, we will assume an RMT of 50% MSO.

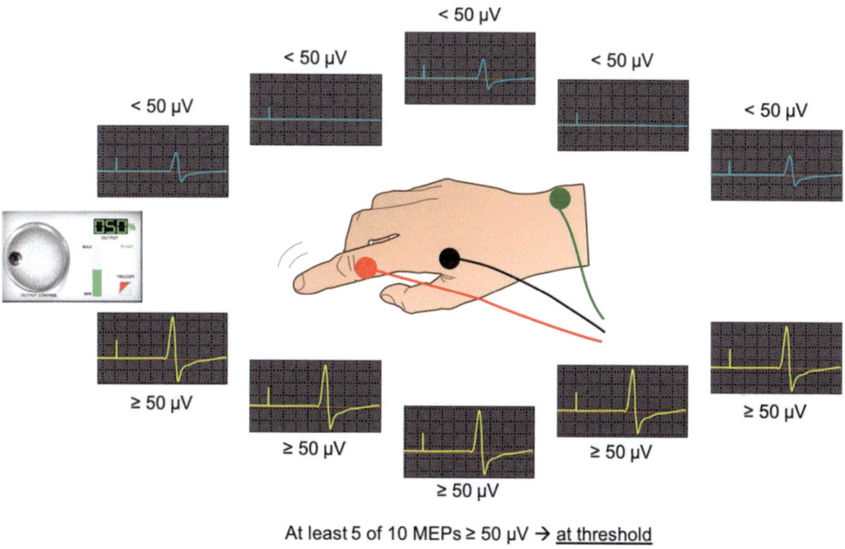

Step 4: Identify the location of the median nerve proximal to the wrist. It should be located along the midline between the flexor tendons. Apply a small amount of conductive gel (or saline) to the anode and cathode of the stimulating electrodes and place them against the skin, lengthwise on the wrist, with the cathode proximal.

Step 5: Give pulses of electrical stimulation at increasing intensities until a visible twitch is observed in the thenar muscles (of the thumb), adjusting the position of the stimulating electrodes as needed. Secure final position in place with medical tape.

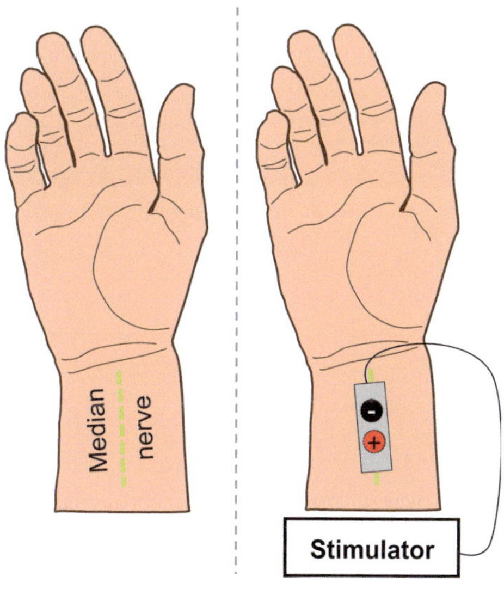

Step 6: With intensity set to 120% RMT (60% MSO in this example), deliver 10 TMS pulses (spaced every 6–10 s). The average MEPs elicited from TMS will serve as the unconditioned (TS alone) response.

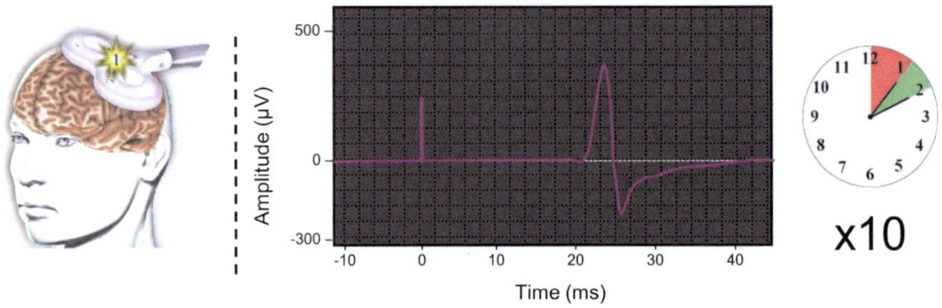

Step 7: Set the TS (TMS) and CS (electrical) intensities. The TMS intensity will be 120% RMT as above (60% MSO in this example), while the electrical stimulus will be at the intensity that elicits a visible twitch of the thenar (Step 5).

Step 8: Choose an ISI from 18 to 25 ms. For this example, we will use a 21-ms ISI.

Step 9: Deliver 10 trials of paired electrical-TMS stimuli (spaced every 6–10 s). Average the MEPs that result from the conditioned TS and express them as the % change (or ratio) of the unconditioned MEPs from Step 6. This will be the SAI response.

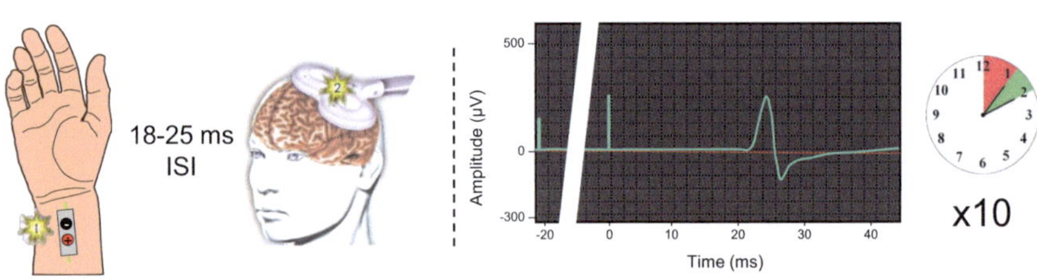

Notes:

– As illustrated in *Sect. 5.1*, a range of ISIs can be used to generate a curve for SAI. This is a form of an I/O curve (*Sect. 4.1*) that evaluates the conditioning effects (y-axis) of a range of ISIs (x-axis).

- A more individualized approach is to set the ISI based on the latency of the N20 somatosensory potential (either with a set ISI or as an I/O curve). This can be recorded using a dedicated EEG cap and amplifiers, but it is also possible to use the same EMG amplifier and two scalp electrodes with the active positioned over the targeted somatosensory cortex and a reference on the frontal midline (Fz in the International 10-20 System for EEG electrode placement). The latency of the N20 can be determined by administering ≥500 electrical stimuli to the median nerve and averaging the somatosensory evoked cortical response (details on this procedure can be found in Clinical Neurophysiology textbooks). The effects of SAI have been observed using ISIs that are ~1–4 ms following the N20 latency with the peak effects occurring 2–3 ms post-N20.
- Given the hypothesized role of intracortical (sensory to motor) interactions in the mechanisms of SAI, the direction of the induced current in the brain is believed to play a significant role. For this reason, we recommend applying SAI using a monophasic waveform inducing a PA current in the brain (*Sect. 1.5*) as there is some evidence that PA currents preferentially activate connections onto M1 from somatosensory cortex.

5.3.3 *Triple-Pulse Stimulation Technique*

TST was developed to allow for quantification of corticospinal tract failure due to axonal loss. This approach is more accurate and reliable than measuring MEP amplitude alone as a suprathreshold TMS pulse to M1 activates only a portion of the descending axons, whereas a suprathreshold electrical stimulus to the PNS activates all axons. The protocol for TST consists of delivering a TMS pulse to M1 and two electrical stimuli to peripheral locations along the motor pathway (usually the ulnar nerve at the wrist and the brachial plexus at Erb's point).

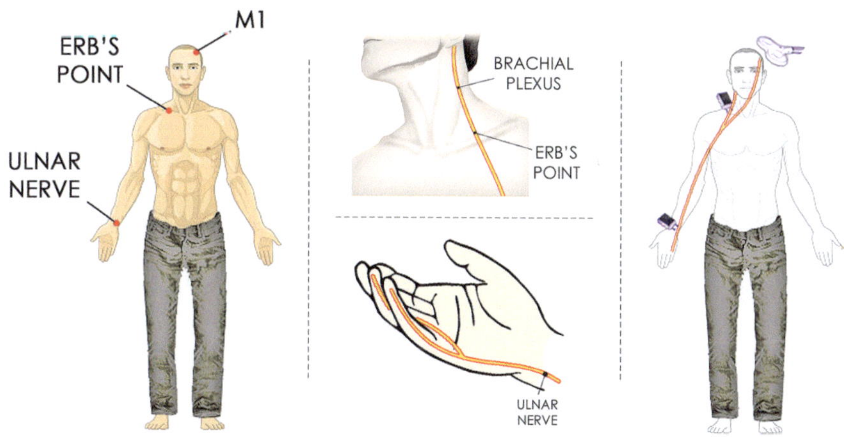

When delivered at specific intervals, the axonal signals from the TMS pulse to M1 "collide" along the motor pathway with those from peripheral electrical stimuli (see figure below). More specifically, a TMS pulse to M1 elicits desynchronized descending axonal signals, which collide with synchronized ascending axonal signals generated by the electrical stimulation of the ulnar nerve at the wrist. Some of the ascending signal remain and collide with descending signals elicited by electrical stimulation of the brachial plexus at Erb's point. The remaining descending signals from Erb's point will be the same axons originally stimulated by the TMS pulse to M1 but will now be synchronized.

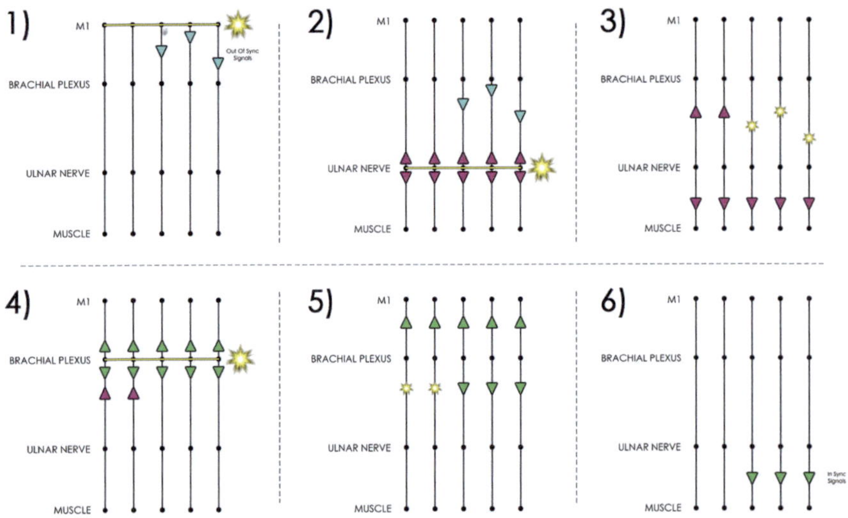

The correct ISI between each stimulus, labeled ISI 1 and ISI 2, is unique to each subject's physiology. ISI 1 is calculated by subtracting the M-wave latency from stimulating the wrist from the MEP latency by stimulating M1. ISI 2 is calculated by subtracting the M-wave latency from stimulating the wrist from the M-wave latency from stimulating Erb's point.

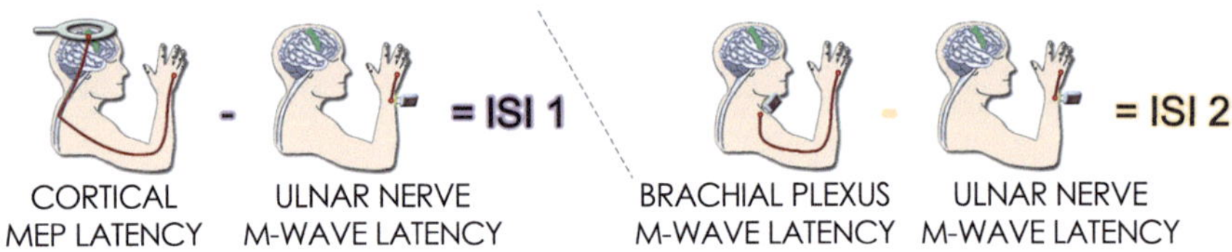

To perform TST, we need to measure two different amplitudes, a TST control amplitude reflecting all the existing axons and a TST test amplitude, reflecting the portion of axons activated by TMS. The TST control amplitude is measured by first stimulating at Erb's point, then the wrist, then again at Erb's point using ISI 2 between all stimuli. The TST test amplitude is measured by first stimulating at M1, then the wrist, then Erb's point using ISI 1 and ISI 2, respectively.

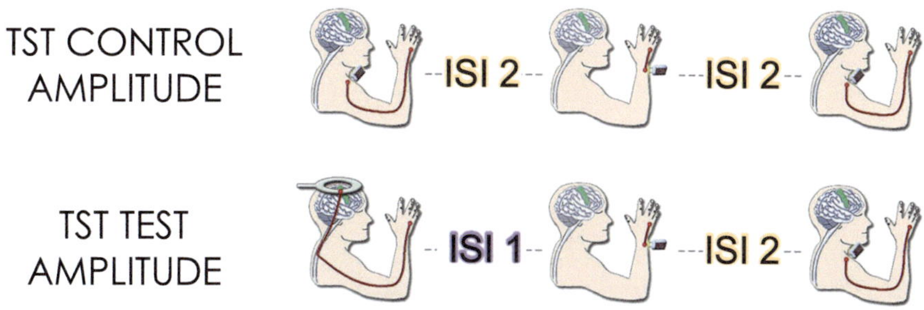

To measure the extent of axon loss in corticospinal tract, we subtract the test amplitude from the control amplitude. The greater the difference, the more axonal loss.

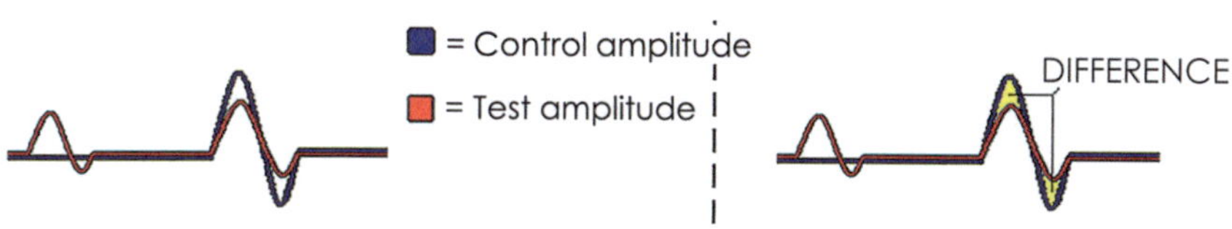

Although this technique can be used in a variety of muscles, we will demonstrate it in the left ADM (*Sect. 3.1*).

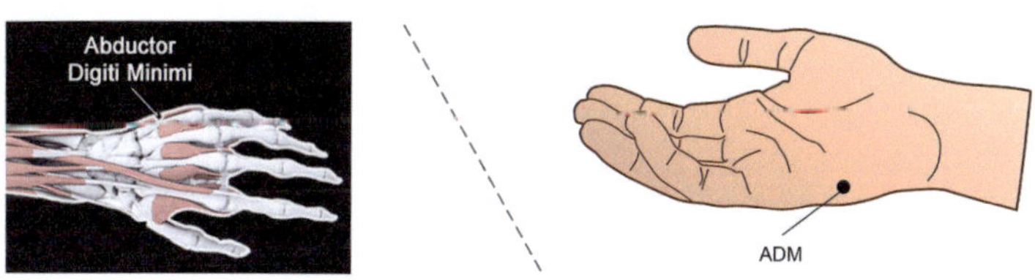

Note This procedure will require a TMS machine, two peripheral electrical stimulators, and a pulse timing mechanism.

5.3.4 Step for Conducting Triple-Pulse Stimulation with the Right ADM as the Target Muscle

Step 1: Attach EMG electrodes to the right ADM using a belly-tendon montage (Sect. 3.1). The ground electrode should be placed on the back of the hand rather than the bony wrist to leave space for the stimulating electrodes over the ulnar nerve in the wrist.

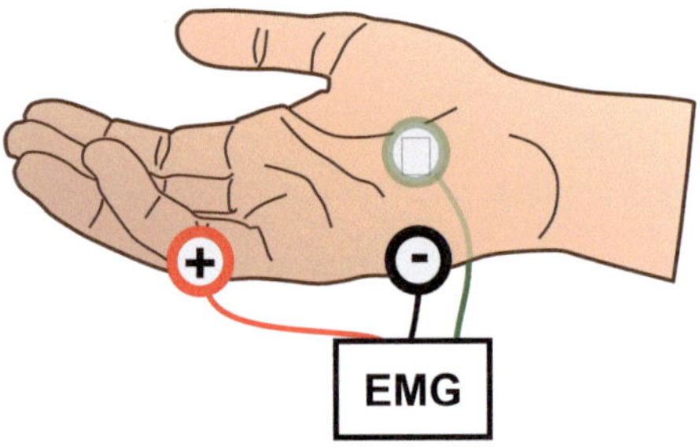

Step 2: Determine the FDI motor hotspot (*Sect. 3.3*). Remember that when determining the hotspot, one should start at a low intensity (i.e., below threshold) and increase until responses are large and consistently observed (i.e., above threshold). The hands should be completely at rest (i.e., no EMG activity) for the hotspot.

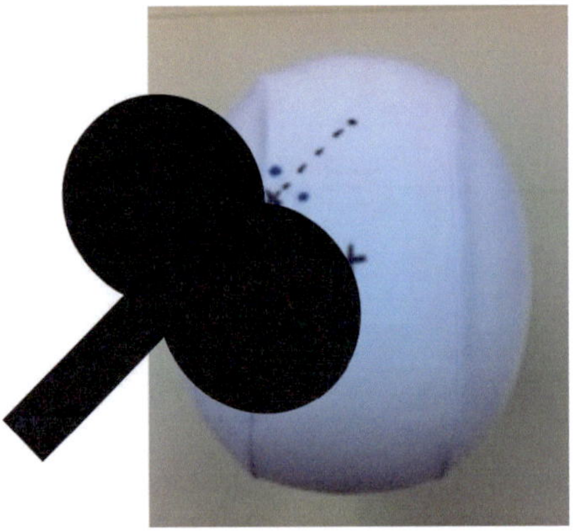

Step 3: Determine the RMT for the ADM hotspot (*Sect. 3.4*). For this example, we will assume an RMT of 50% MSO.

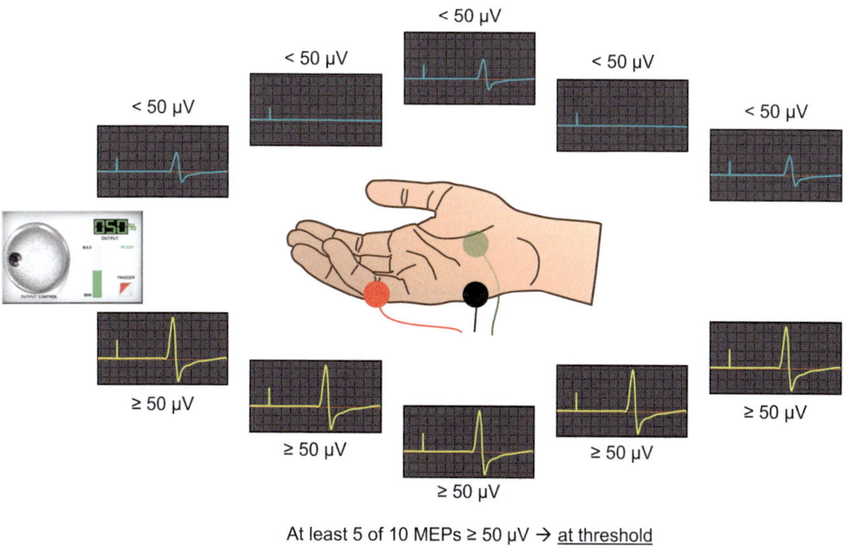

Step 4: Identify the location of the ulnar nerve proximal to the pisiform bone of the wrist. It should be located along the edge of the forearm. Apply a small amount of conductive gel (or saline) to the anode and cathode of the stimulating electrodes and place them against the skin, lengthwise on the wrist, with the cathode proximal.

Step 5: Give pulses of electrical stimulation at increasing intensities until a visible twitch is observed in the ADM, adjusting the position of the stimulating electrodes as needed. Secure final position in place with medical tape.

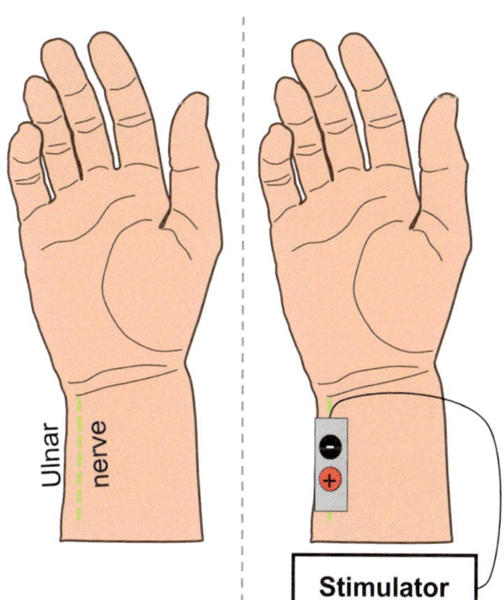

Step 6: Locate Erb's point 2–3 cm above the clavicle. Apply a small amount of conductive gel (or saline) to the anode and cathode of the stimulating electrodes and place them against the skin. Place the cathode at Erb's point and the anode lateral toward the shoulder. Secure in place with medical tape.

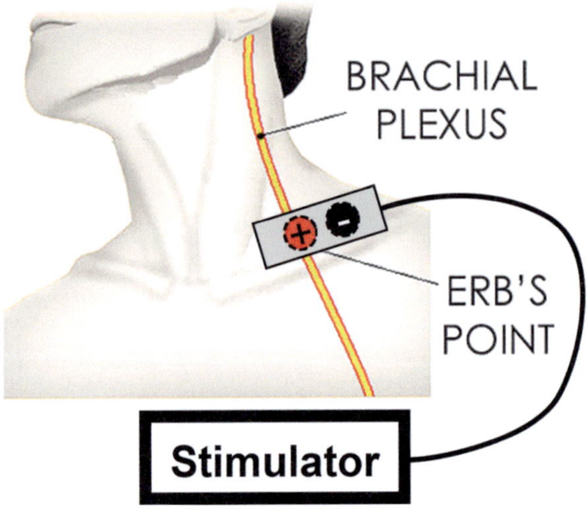

Step 7: Link TMS to electrical stimulators via a precise timing mechanism/trigger.

Step 8: Set intensity to 120% RMT (in this example, 60% MSO). Deliver 10 TMS pulses (spaced 6–10 s apart to avoid cumulative effects) to the ADM motor hotspot. For each trial, note both the amplitude and latency of the MEP (*Sect. 3.3*).

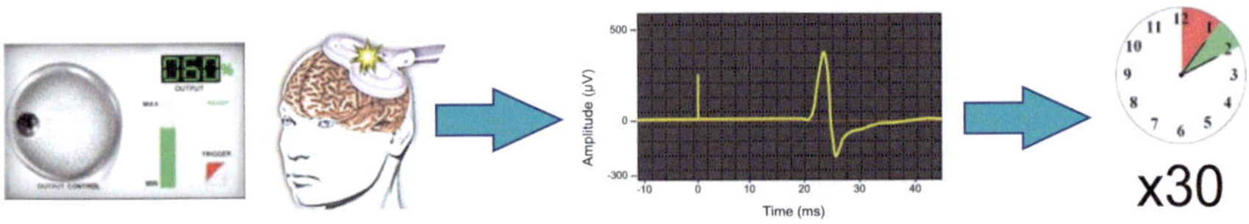

Step 9: Determine the shortest latency to the MEP onset in the series. This *time to MEP* value will be labeled value "A." For this example, we will assume a latency of 19 ms.

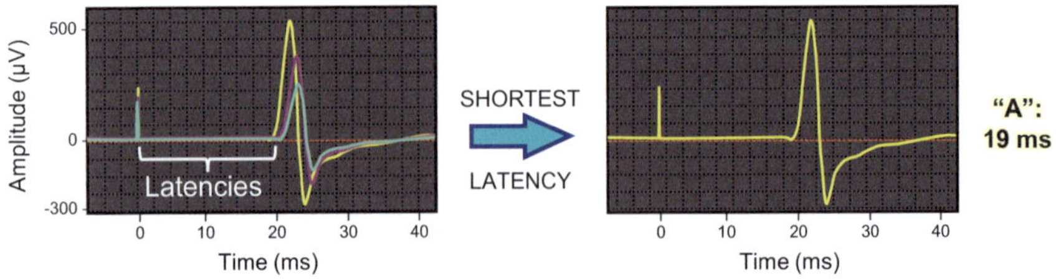

Step 10: Deliver electrical stimuli to the ulnar nerve at increasing intensities, measuring the resulting M-wave amplitude, until reaching the *supramaximal* intensity (when the M-wave does not get any larger).

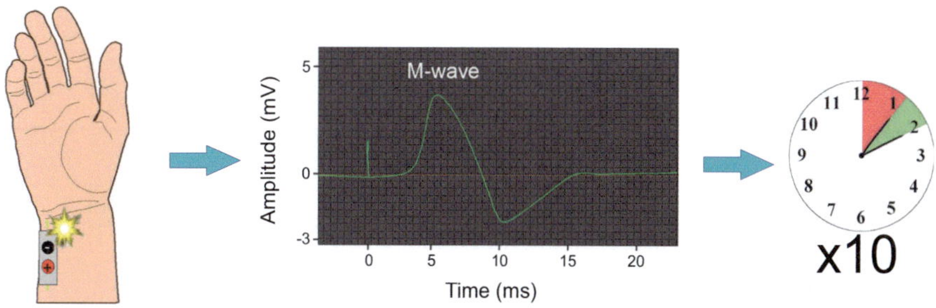

Step 11: Apply 10 electrical stimuli to the ulnar nerve at the supramaximal intensity from Step 10. Determine the shortest latency to the M-wave onset in the series. This *time to M-wave* value will be labeled value "B." For this example, we will assume a latency of 3 ms.

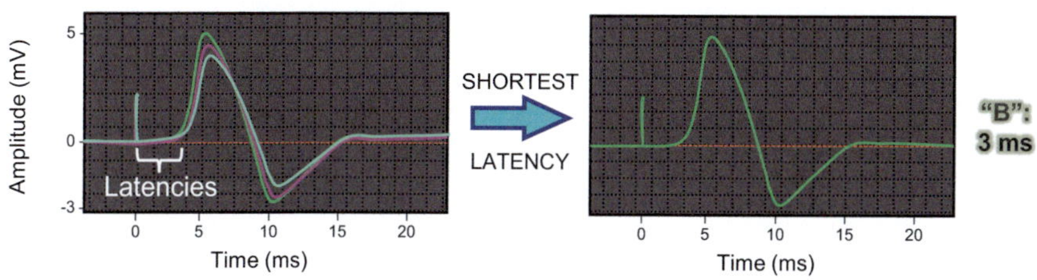

Step 12: Determine the supramaximal intensity for Erb's point following the same procedure as in Step 10. The supramaximal intensity for Erb's point should be at or above that of the wrist so the intensity found in Step 10 can be used as a starting point.

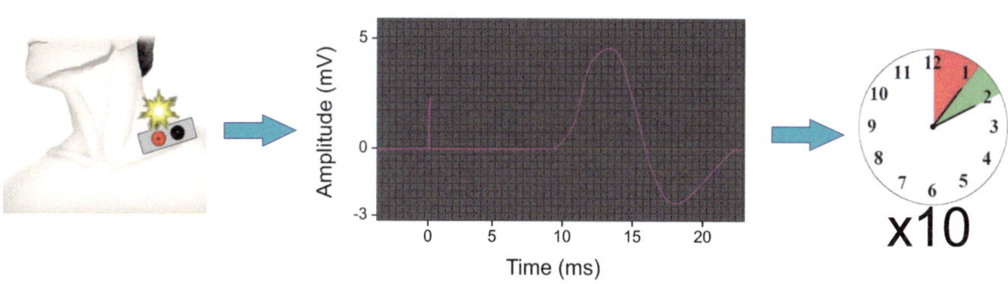

Step 13: Apply 10 electrical stimuli to Erb's point at the supramaximal intensity from Step 13. Determine the shortest latency to the M-wave onset in the series. This *time to M-wave* value will be labeled value "C." For this example, we will assume a latency of 10 ms.

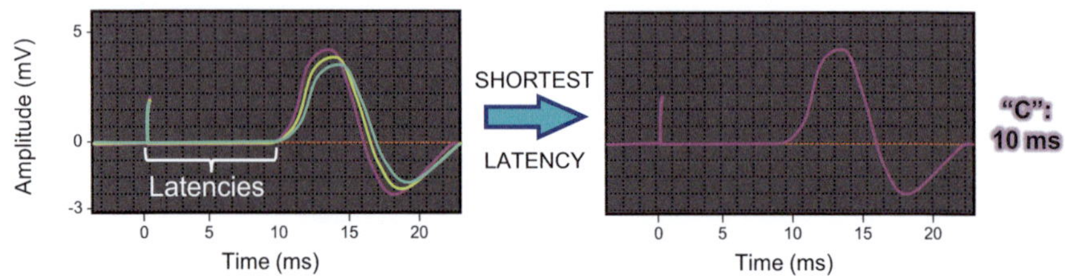

Step 14: Determine ISI 1 by subtracting value "B" from value "A." For this example, ISI 1 would equal 16 ms.

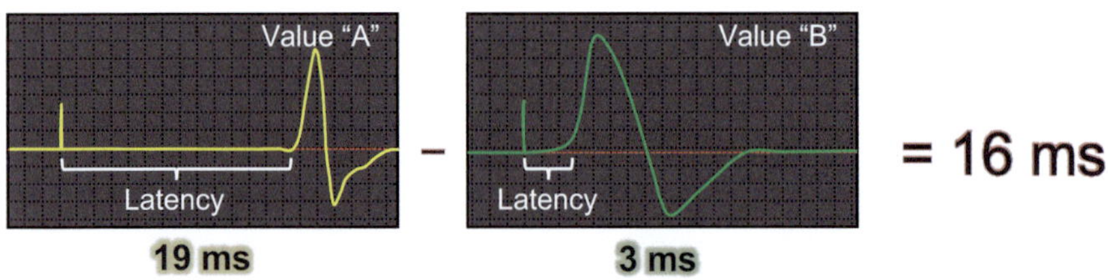

Step 15: Determine ISI 2 by subtracting value "B" from value "C." For this example, ISI 2 would equal 7 ms.

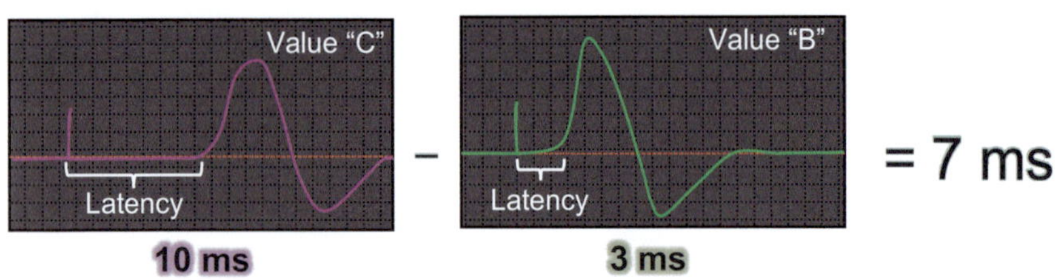

Step 16: Set TMS intensity to 120% RMT (60% MSO in this example) and the electrical stimuli to the supramaximal intensities found in Steps 10 and 12.

Step 17: Set timing mechanism to trigger as follows:

- *M1 TMS pulse*
- *ISI 1*
- *Wrist electrical pulse*
- *ISI 2*
- *Erb's point electrical pulse*

Step 18: Apply 10 trials of the above sequence (spaced 6–10 s apart). Measure the amplitude of the evoked response for each trial.

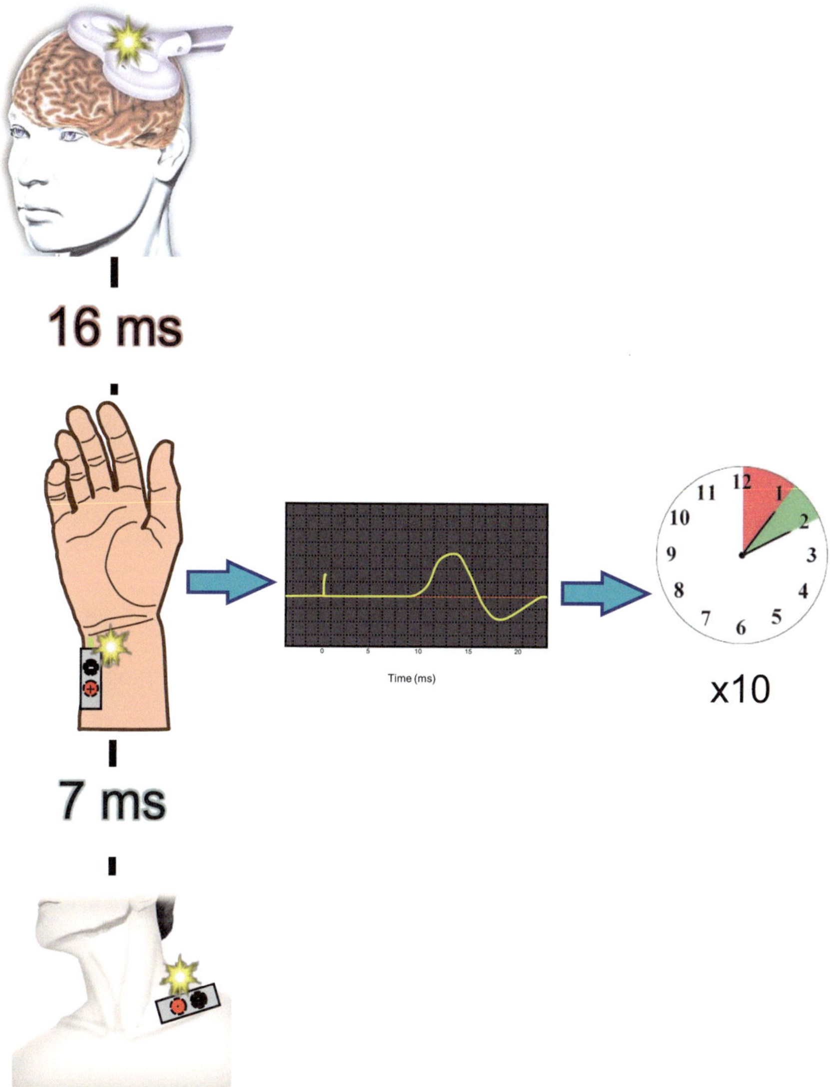

Step 19: Average the amplitude of the 10 trials from Step 18. This will be your TST *test* amplitude value.

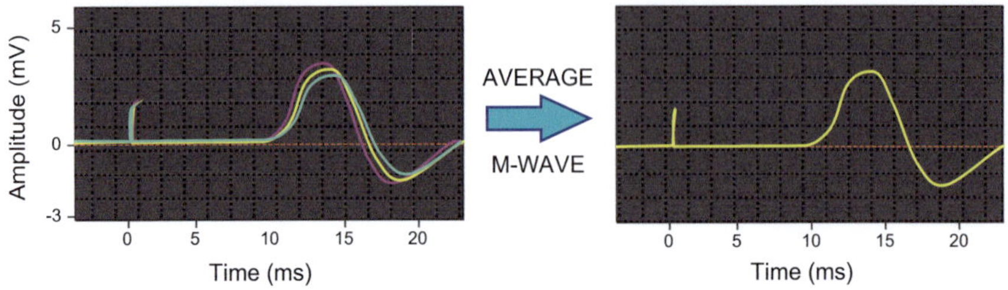

Step 20: Set the electrical stimuli to the supramaximal intensities found in Steps 10 and 12.

Step 21: Set timing mechanism to trigger as follows:

- *Erb's point electrical pulse*
- *ISI 2*
- *Wrist electrical pulse*
- *ISI 2*
- *Erb's point electrical pulse*

Step 22: Apply 10 trials of the above sequence (spaced 6–10 s apart). Measure the amplitude of the evoked response for each trial.

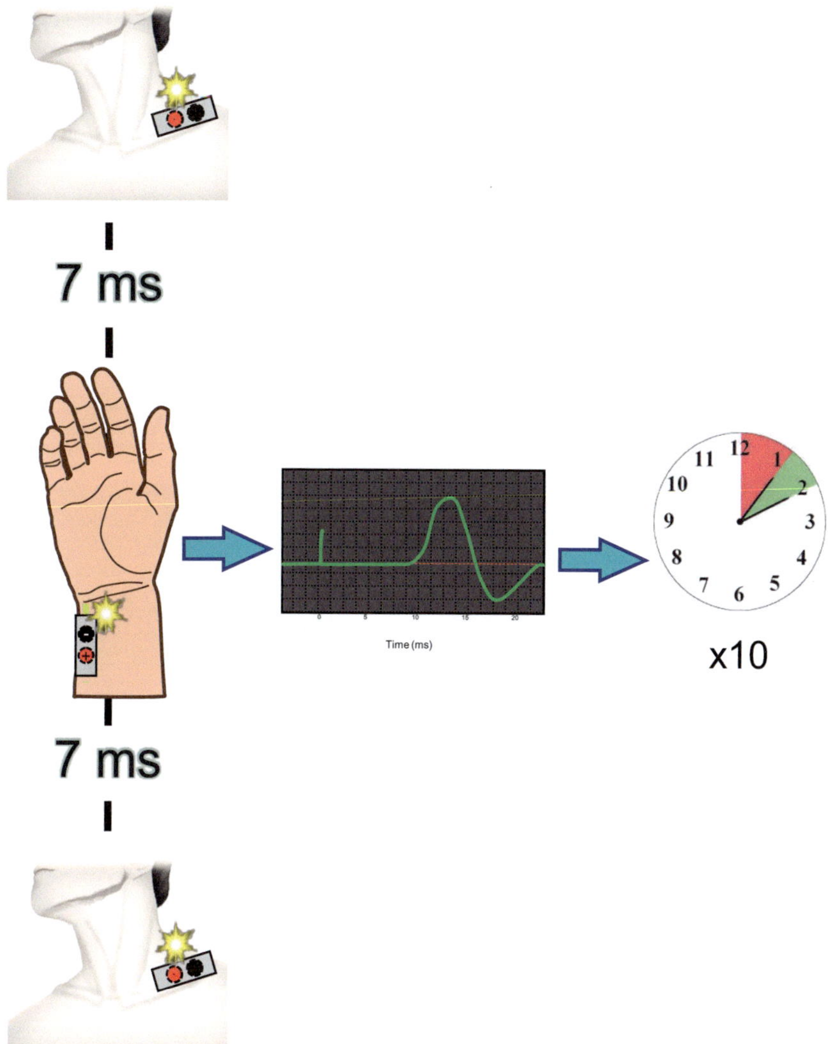

Step 23: Average the amplitude of the 10 trials from Step 22. This will be your TST *control* amplitude value.

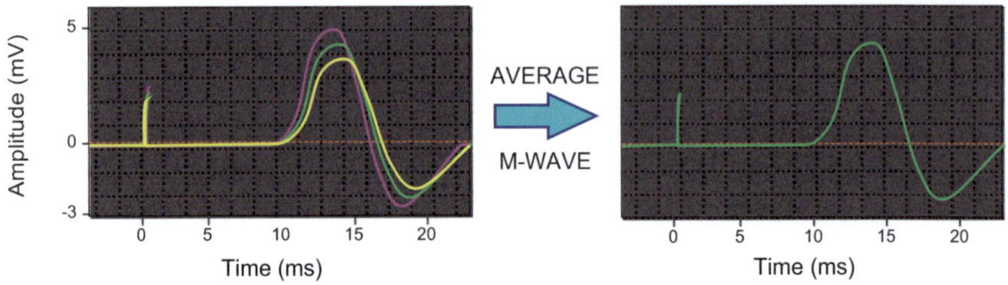

Step 24: Subtract TST test curve value (Step 19) from TST control curve value (Step 23).

- **The difference between the TST test curve and TST control curve is a measure of corticospinal axonal loss.**
- **A greater difference indicates worse degeneration.**

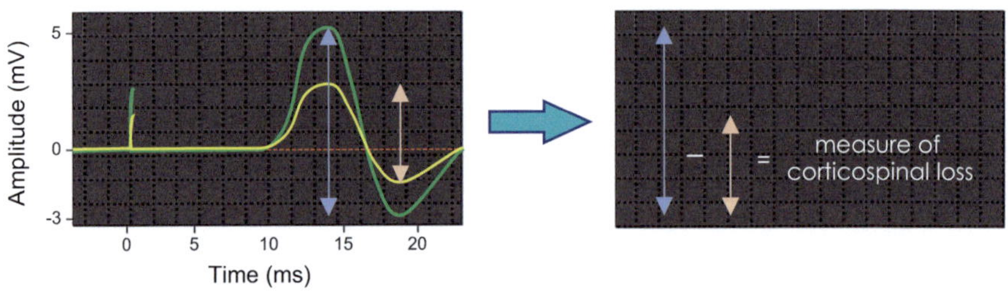

Notes:

- The area under the curve can be a more accurate method to calculate evoked activity in cases of neurodegeneration.

Further Reading

Section 5.1

Maeda, F., Gangitano, M., Thall, M., & Pascual-Leone, A. (2002). Inter-and intra-individual variability of paired-pulse curves with transcranial magnetic stimulation (TMS). Clinical Neurophysiology, 113(3), 376-382.

Nakamura, H., Kitagawa, H., Kawaguchi, Y., & Tsuji, H. (1997). Intracortical facilitation and inhibition after transcranial magnetic stimulation in conscious humans. The Journal of Physiology, 498(Pt 3), 817-823.

Valls-Solé, J., Pascual-Leone, A., Wassermann, E. M., & Hallett, M. (1992). Human motor evoked responses to paired transcranial magnetic stimuli. Electroencephalography and Clinical Neurophysiology/Evoked Potentials Section, 85(6), 355-364.

Section 5.2

Boroojerdi, B., Foltys, H., & Meincke, U. (1999). Transcallosal inhibition and motor conduction studies in patients with schizophrenia using transcranial magnetic stimulation. The British Journal of Psychiatry, 175(4), 375-379.

Ferbert, A., Priori, A., Rothwell, J.C., Day, B.L., Colebatch, J.C., & Mardsen, C.D. (1992). Interhemispheric Inhibition of the human Motor Cortex. Journal of Physiology, 453, pp. 525-546.

Kobayashi, M., & Pascual-Leone, A. (2003). Transcranial magnetic stimulation in neurology. The Lancet Neurology, 2(3), 145-156.

Koch, G., Franca, M., Mochizuki, H., Marconi, B., Caltagirone, C., & Rothwell, J. C. (2007). Interactions between pairs of transcranial magnetic stimuli over the human left dorsal premotor cortex differ from those seen in primary motor cortex. The Journal of Physiology, 578(2), 551-562.

Ni, Z., Leodori, G., Vial, F., Zhang, Y., Avram, A. V., Pajevic, S., Basser, P. J. & Hallett, M. Measuring latency distribution of transcallosal fibers using transcranial magnetic stimulation. Brain Stimulation 13, 1453–1460 (2020).

Section 5.3

Bühler, R., Magistris, M. R., Truffert, A., Hess, C. W., & Rösler, K. M. (2001). The triple stimulation technique to study central motor conduction to the lower limbs. Clinical Neurophysiology, 112(5), 938-949.

Di Lazzaro, V., Oliviero, A., Profice, P., Pennisi, M. A., Di Giovanni, S., Zito, G., Tonali, P. & Rothwell, J. C. Muscarinic receptor blockade has differential effects on the excitability of intracortical circuits in the human motor cortex. Exp Brain Res 135, 455–461 (2000).

Komissarow, L., Rollnik, J. D., Bogdanova, D., Krampfl, K., Khabirov, F. A., Kossev, A., ... & Bufler, J. (2004). Triple stimulation technique (TST) in amyotrophic lateral sclerosis. Clinical Neurophysiology, 115(2), 356-360.

Tokimura, H., Di Lazzaro, V., Tokimura, Y., Oliviero, A., Profice, P., Insola, A., Mazzone, P., Tonali, P. & Rothwell, J. C. Short latency inhibition of human hand motor cortex by somatosensory input from the hand. J. Physiol. (Lond.) 523 Pt 2, 503–513 (2000).

Chapter 6
Neuromodulation

Abstract While single TMS pulses have short-lived effects on local and distributed physiology, repeated pulses in frequency-specific patterned trains can induce more durable effects on physiology and behavior, well outlasting the stimulation itself. This phenomenon, known as neuromodulation, can be used to study brain function in health and disease, and forms the basis of many clinical treatment applications. Neuromodulation is likely a result of engaging aspects of neuroplasticity. As in pharmacology, where a drug can have different effects depending on its chemical composition, rTMS stimulation parameters critically influence the resulting neuromodulatory effects on the brain. This chapter outlines several key approaches with established protocols and the most used parameters from the literature. These approaches can be used to evaluate the physiological effects of rTMS (often referred to as cortical plasticity) using either a TMS brain target alone or a TMS brain target paired with a precisely timed peripheral nerve stimulation. These same approaches can be adapted for use as behavioral or therapeutic interventions. Scenarios for lab-based investigation and clinical treatment application are discussed.

D. J. Edwards et al., *A Practical Manual for Transcranial Magnetic Stimulation*, https://doi.org/10.1007/978-3-031-62304-2_6

6.1 Repetitive TMS and Evaluation of the Mechanisms of Cortical Plasticity

rTMS paradigms rely on patterned trains of pulses to induce cortical effects that outlast the duration of stimulation. Clinically, rTMS can be used to normalize aberrant patterns of focal and network cortical activity, which may lead to alleviation of disease state symptoms (see *Sect. 6.3* for more details). Additionally, the presumed rTMS neuromodulatory effects can be utilized to investigate causal brain–behavior relationships by evaluating changes in performance on a particular task after applying rTMS.

As with paired-pulse TMS (*Sect. 5.1*), there are a number of rTMS protocols that can be used to induce varying changes in neural activity of the stimulated region and associated networks. Historically, rTMS has been applied as either a continuous low-frequency train (~1 Hz) or a pattern of intermittent on–off high-frequency trains (≥10 Hz).

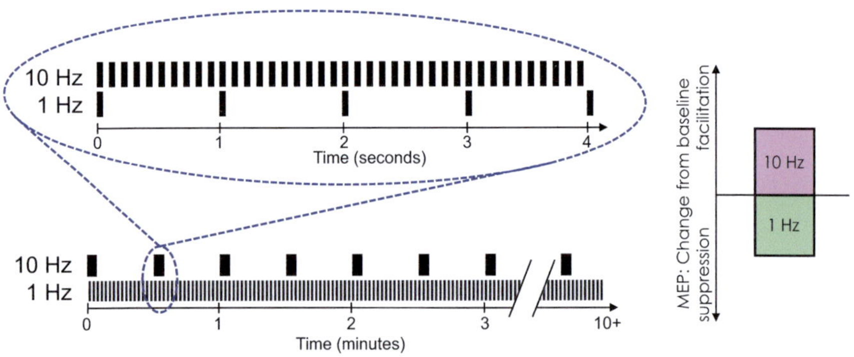

More recently, TBS has been developed as an ultra-high frequency pattern of rTMS. The most common applications of TBS consist of applying bursts of three pulses at 50 Hz that are repeated at 5 Hz. This coupling of gamma and theta frequencies in TBS was patterned after endogenous cerebral oscillations associated with neural plasticity. Like other forms of rTMS, TBS can be applied continuously (cTBS) or intermittently (iTBS) to induce effects that resemble the synaptic mechanisms of long-term depression (LTD) and long-term potentiation (LTP), respectively. TBS protocols are thought to be more efficient than conventional rTMS, yielding aftereffects that can last 30 or more minutes from only 600 pulses.

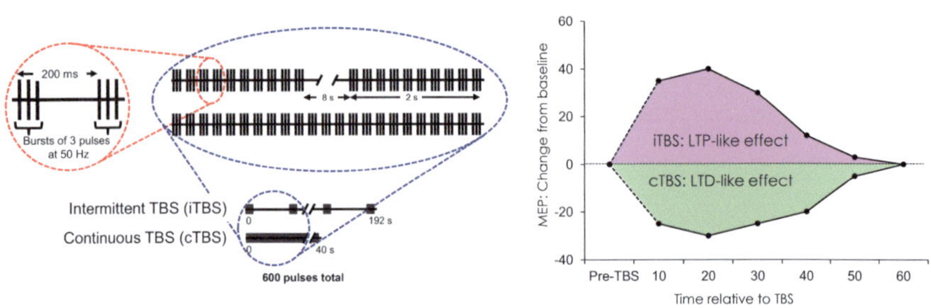

rTMS parameters define and distinguish each protocol. Historically, these parameters have not always been standardized and there may be important reasons (e.g., device limitation or study design) to make adjustments. Attempts have been made to standardize certain rTMS protocols, especially those used for clinical purposes (see *Sect. 6.3*). The most commonly used parameters are summarized in the table below.

Frequency	Protocol	Trial unit	Inter-pulse interval	Pulses/burst	Inter-burst interval	Pulses/train	Train duration	Inter-train interval	# Trains	Total # pulses	Time
Low	1 Hz	Pulse	1000 ms	n/a	n/a	10	10 s	1 s	160	1600	26.67 min
High	10 Hz	Pulse	100 ms	n/a	n/a	40	4 s	26 s	75	3000	37.5 min
	20 Hz	Pulse	50 ms	n/a	n/a	40	2 s	28 s	40	1600	20 min
Ultra-high	cTBS	Burst	20 ms	3	200 ms	30	2 s	0.2 s	20	600	40 s
	iTBS	Burst	20 ms	3	200 ms	30	2 s	8 s	20	600	192 s

One critical parameter that is not included above is stimulation intensity. While there are safety guidelines governing maximum intensities for different frequencies (*Sect. 2.2*), intensity is often very specific to the targeted region and the application (i.e., clinical vs. non-clinical) and can vary between 70% and 120% RMT, while some protocols have historically used 80% AMT.

Two other important factors to consider are stimulation target and outcome measure. These two factors are interrelated and depend on the objective of the study or treatment. Most of the presumed effects attributed to different rTMS protocols are based on observations of their effects on MEP amplitudes elicited by single-pulse TMS to M1. For example, 1 Hz and cTBS are presumed to be "inhibitory," while 10 Hz, 20 Hz, and iTBS are presumed to be "facilitatory" based on decreases or increases, respectively, in MEP amplitude. Moreover, even these effects are based on group-level averages and there can be considerable inter- and intra-individual variability in the response to a given protocol (see *Further Reading* for more information). For stimulation outside M1, the effects of a given protocol are often assumed to follow the same direction as with M1. However, it is important to be aware of such assumptions when there is not an independent physiologic measure of the effect, especially when interpreting changes in mood or behavioral measures. The mechanisms of rTMS in non-motor brain regions can be also studied with neuroimaging modalities such as EEG, positron emission tomography (PET), and functional MRI.

Pharmacological studies can also provide mechanistic insight into the effects of rTMS on neural activity. For example, administration of NMDA receptor antagonists, such as memantine, has been shown to blunt the aftereffects of TBS, leading to the interpretation that its mechanisms are NMDA receptor dependent.

As with other TMS protocols, the effects of rTMS can be altered in certain neurologic and psychiatric conditions. Accordingly, the efficacy of specific rTMS protocols can provide important information on disease pathophysiology. For example, in individuals with migraine, the ability of 1 Hz M1 rTMS to suppress MEP amplitudes has been shown to be reduced compared to controls, putatively reflecting altered cortical excitability and neurotransmitter function. In Alzheimer's disease, the effects of iTBS on MEPs have been shown to be similarly blunted, while cTBS effects are spared, or even enhanced, possibly reflecting the impact of amyloid-β and other neurodegenerative processes on glutamatergic function. By contrast, the effects of both cTBS and iTBS on MEPs have been shown to be greater in individuals with autism spectrum disorder, which is speculated to reflect a neural state of hyper-plasticity.

Similar to the conditioning effects of paired-pulse TMS and paired TMS-electrical stimulation discussed in Chap. 5, the neuromodulatory response of rTMS protocols (including cTBS and iTBS) on MEP amplitude can be expressed as either (1) the % change from a pre-rTMS baseline or (2) a ratio of the pre-rTMS baseline:

$$(1)\ \frac{(\textbf{post-rTMS MEP} - \textbf{pre-rTMS MEP})^{*}\,\textbf{100}}{(\textbf{pre-rTMS MEP})} \qquad (2)\ \frac{\textbf{post-rTMS MEP}}{\textbf{pre-rTMS MEP}}$$

These formulae can also be adapted to express the effects of rTMS on other physiologic measures (i.e., TMS-evoked EEG potentials) or behavioral changes.

Below, we will demonstrate rTMS effects on M1 excitability using the iTBS protocol and discuss the modifications necessary to perform other rTMS protocols.

Note As the pulse trains are generated in rapid succession through a single coil, this procedure may require a TMS machine with multiple charging units and an air- or fluid-based cooling system to prevent the coil from overheating.

6.1.1 Steps for Evaluating Cortical Plasticity in Left M1 with the iTBS Protocol

Step 1: Attach EMG electrodes to the right FDI using a belly-tendon montage with a ground on the bony wrist (*Sect. 3.1*).

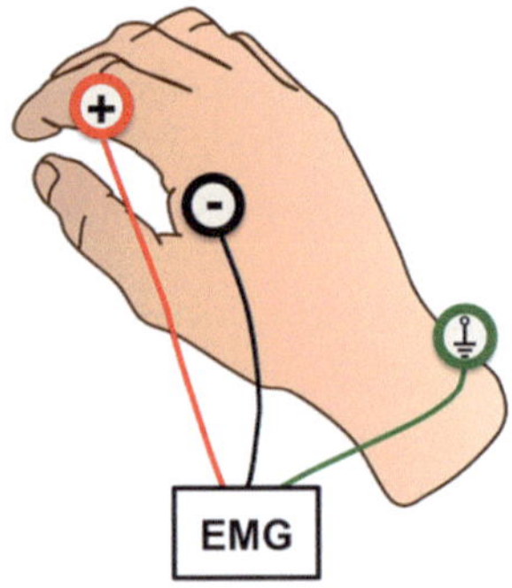

Step 2: Determine the FDI motor hotspot (*Sect. 3.3*). Remember that when determining the hotspot, one should start at a low intensity (i.e., below threshold) and increase until responses are large and consistently observed (i.e., above threshold). The hands should be completely at rest (i.e., no EMG activity) for the hotspot.

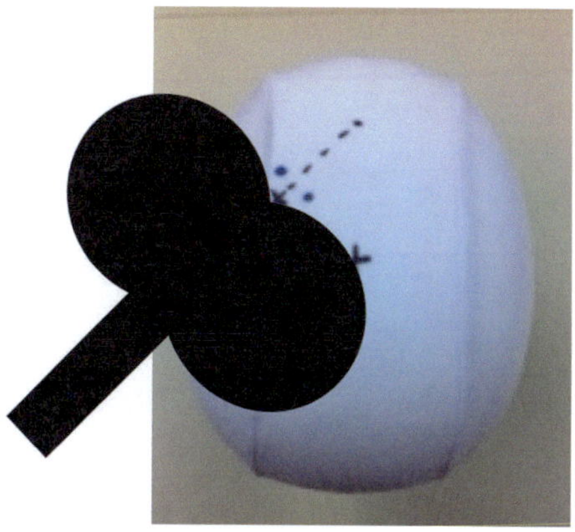

Step 3: Determine the RMT for the FDI hotspot (*Sect. 3.4*). For this example, we will assume an RMT of 50% MSO.

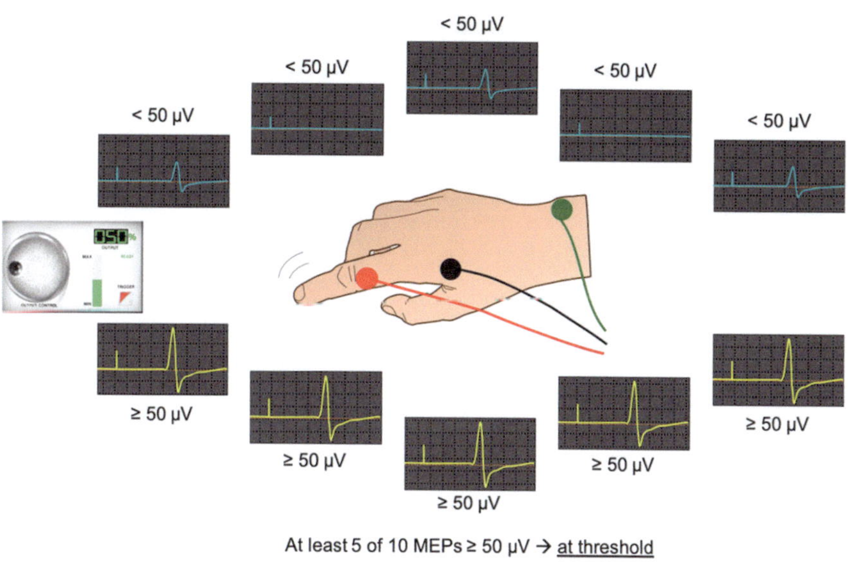

Step 4: Instruct the participant to keep both of his/her hands at rest for the remainder of the session (especially the time before, during, and after the TBS protocol).

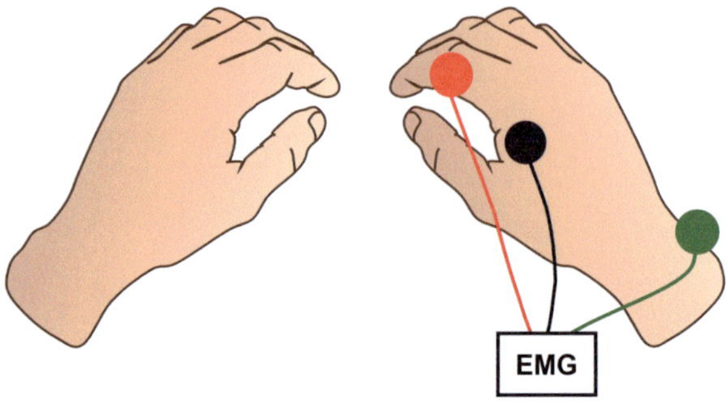

Step 5: Set intensity to 120% of the RMT from Step 3 (60% MSO in this example). Deliver 30 single pulses (spaced every 6–10 s). The average MEP amplitude will serve as the baseline (pre-TBS) response.

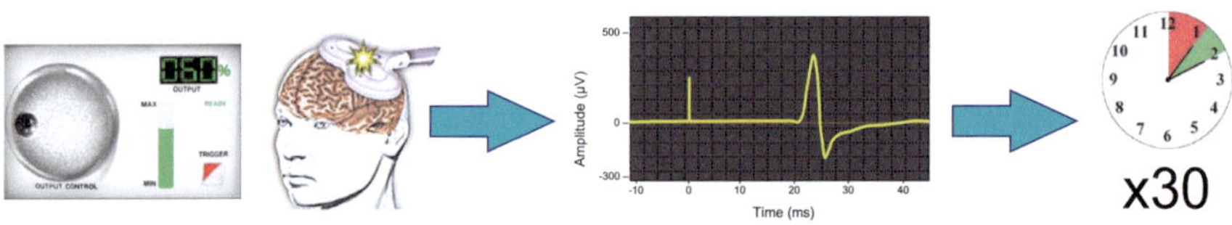

Step 6: Program the stimulator to deliver iTBS with 600 total pulses (using the introductory table for a reference of parameters).

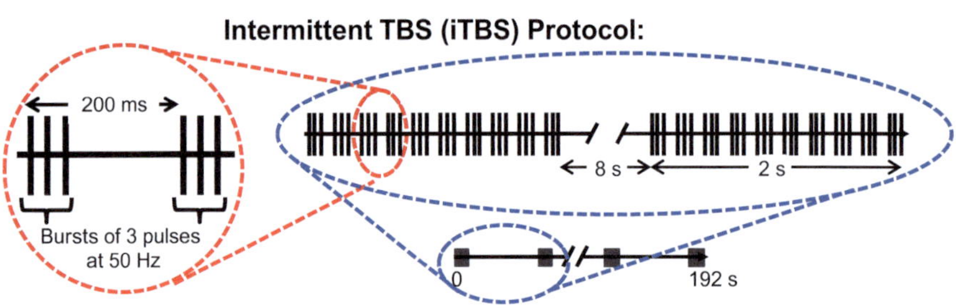

Step 7: Set intensity to 70% RMT found in Step 4 (35% MSO in this example).

Step 8: Repeat Step 5 for all post-iTBS time-points. In this example, we are measuring the average MEP amplitude every 10 min post-iTBS for up to 1 h.

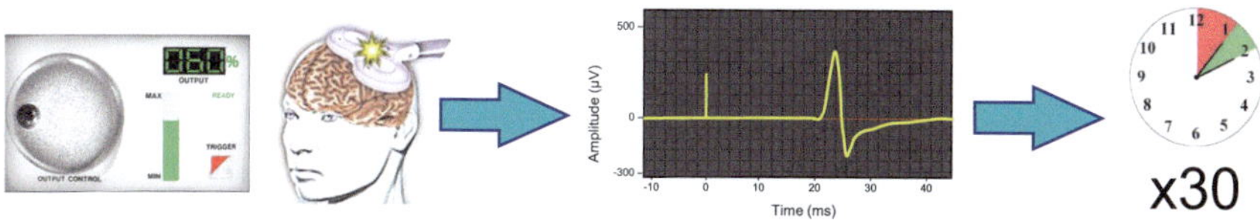

Step 9: Express the amplitude of MEPs from each post-iTBS time-point as the % change (or ratio) of the pre-iTBS MEPs from Step 5.

Step 10: Plot the values from Step 9 to form a curve representing the neuromodulatory response. In this example, the positive change from baseline reflects an LTP-like response.

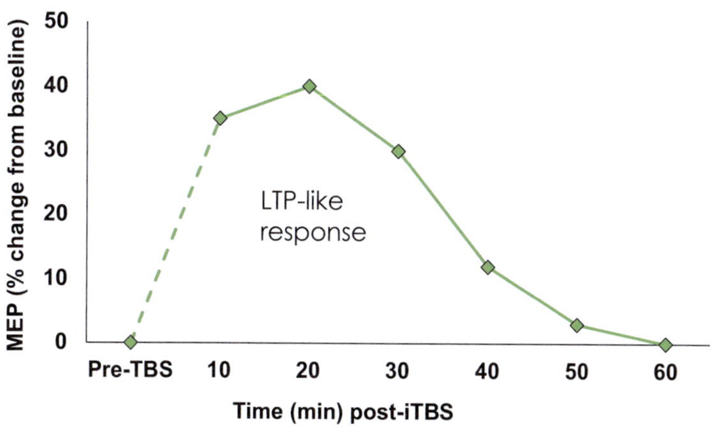

Notes:

- A number of studies have shown that the effects of TBS are sensitive to the activity of the target muscle (presumably through changes in the state of the corticospinal pathway). Accordingly, it is advised that participants keep their hands at rest before, during, and after the TBS protocol.
- To improve the stability of the pre- and post-rTMS MEP measurements (Steps 5 and 8), it may be preferential to increase the number of trials per block to at least 25.
- These same steps can be performed to evaluate the effects of any rTMS protocol. Replace iTBS in Step 6 with the appropriate timing parameters from the introductory table. In addition, the intensity in Step 7 may need to be adjusted for the particular protocol.

- The response to rTMS can be influenced by state (e.g., arousal, anxiety, CNS-active medication, hormones) and genetics (e.g., common polymorphisms of CNS-active genes) and by neuropsychiatric diseases. These factors may alter the efficacy and increase the inter- or intra-individual variability in the response.
- rTMS protocols can be used to modify changes in other TMS-based neurophysiologic measures (e.g., IO curves, paired-pulse protocols, and cortical silent period) as well as in other neuroimaging modalities (e.g., PET, functional MRI, EEG).
- MEP amplitudes and other neurophysiologic measures can drift or change over time independent of any particular intervention. Accordingly, comparing active rTMS with a placebo control (i.e., by using a sham TMS coil) can provide a more accurate assessment of the mechanisms of plasticity by controlling for non-specific effects.

6.2 Paired-Associative Stimulation

As discussed in *Sect. 5.3* on SAI, the MEP response to a TMS pulse over M1 can be transiently conditioned by preceding it with an electrical stimulus delivered to a peripheral nerve. When this pair (peripheral electrical + M1 TMS) is repeated at a set frequency, this can induce lasting modulation of corticospinal excitability, reflected as a change in the MEP response to M1 single-pulse TMS alone. Depending on the ISI between the electrical stimulus and the TMS pulse, the excitability of the targeted corticospinal projections (measured as an MEP in the contralateral limb muscles) can be enhanced or diminished for up to ~60 min following the end of the repetitive paired stimulation.

In this protocol, termed paired-associative stimulation (PAS), longer ISIs (17–25 ms; PAS25) typically lead to facilitation (i.e., larger MEPs), whereas shorter ISIs (5–15 ms; PAS10) typically lead to inhibition (i.e., smaller MEPs). In both PAS25 and PAS10 protocols, the conventional repetition rate of the pair is the same (1 pair every 20 s or 0.05 Hz), and it is typically delivered for 30 min (90 pairs). While these are the most common parameters, other repetition rates and durations have been investigated. The figure below shows the general schema of both PAS25 and PAS10.

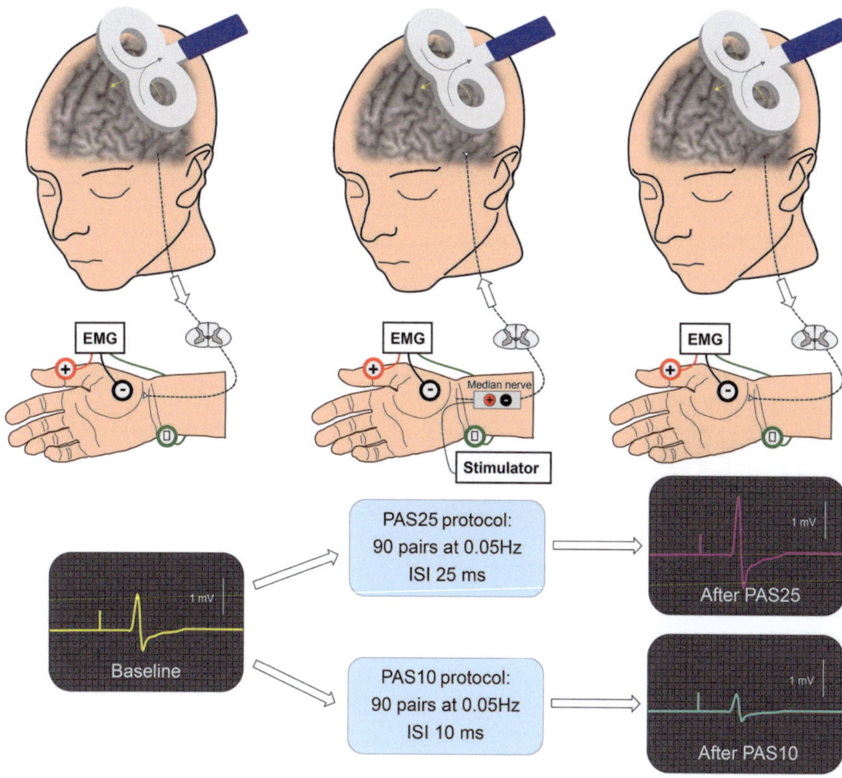

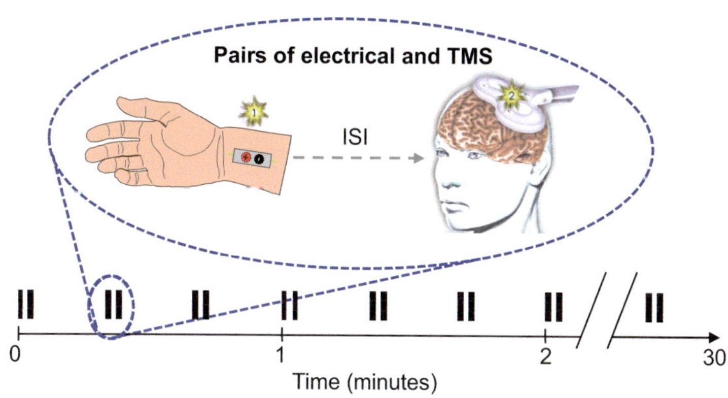

The magnitude and time course of these LTP-like and LTD-like changes are similar to those observed with rTMS (*Sect. 6.1*); however, they are believed to result from engaging mechanisms of spike-timing-dependent plasticity, or the time-specific "association" between the pairing of electrical stimuli and TMS pulses. As with rTMS, pharmacological studies can provide mechanistic insight into the effects of PAS on neural activity. For example, administration of dextromethorphan, an NMDA receptor antagonist, has been shown to block the aftereffects of PAS25, leading to the interpretation that its mechanisms are NMDA receptor dependent, similar to what has been observed with TBS protocols of rTMS (*Sect. 6.1*).

PAS protocols are increasingly being explored in various neurologic and psychiatric patient populations to provide insight into neurophysiologic changes and as a potential therapeutic intervention. For example, in patients with Parkinson's disease (PD), a number of studies have used PAS to demonstrate LTP-like plasticity may be altered in PD patients when they are off medication and may be restored with medication. Another common application is in post-stroke patients, where PAS has been used to reveal how cortical plasticity may change between the sub-acute and chronic phase. PAS also shows potential as a therapeutic intervention to promote motor recovery in post-stroke patients.

As with rTMS (*Sect. 6.1*) and other paired TMS protocols discussed in Chap. 5, the modulatory effects of PAS in M1 can be expressed as either (1) the % change from a pre-PAS baseline or (2) a ratio of the pre-PAS baseline:

$$(1)\ \frac{(\textbf{post-PAS MEP} - \textbf{pre-PAS MEP})^{*}\,\textbf{100}}{(\textbf{pre-PAS MEP})} \qquad (2)\ \frac{\textbf{post-PAS MEP}}{\textbf{pre-PAS MEP}}$$

These formulae can also be adapted to express the effects of PAS on other physiologic measures (i.e., TMS-evoked EEG potentials) or behavioral changes.

Below, we will demonstrate PAS effects on left M1 excitability (targeting the right APB and median nerve) using the PAS25 protocol. We will also discuss how to modify the approach to perform other PAS protocols.

Note This procedure will require a TMS machine, a peripheral electrical stimulator, and a pulse timing mechanism to connect them.

6.2.1 *Steps for Evaluating Changes in Cortical Excitability of Left M1 Using the PAS25 Protocol*

Step 1: Attach EMG electrodes to the right APB using a belly-tendon montage with a ground on the bony wrist (*Sect. 3.1*).

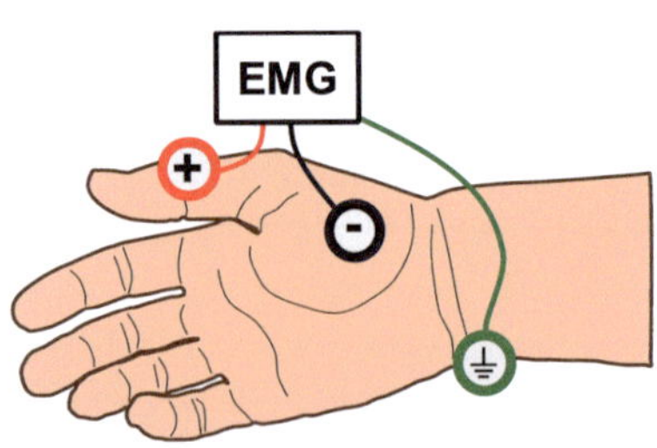

Step 2: Determine the APB motor hotspot (*Sect. 3.3*). Remember that when determining the hotspot, one should start at a low intensity (i.e., below threshold) and increase until responses are large and consistently observed (i.e., above threshold). The hands should be completely at rest (i.e., no EMG activity) for the hotspot.

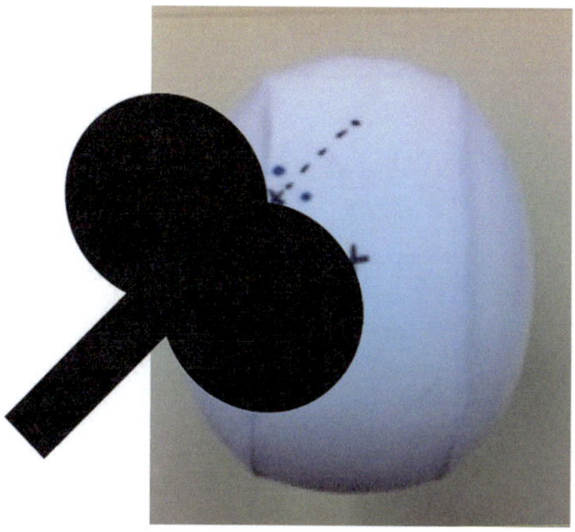

Step 3: Determine the RMT for the APB hotspot (*Sect. 3.4*). For this example, we will assume an RMT of 50% MSO.

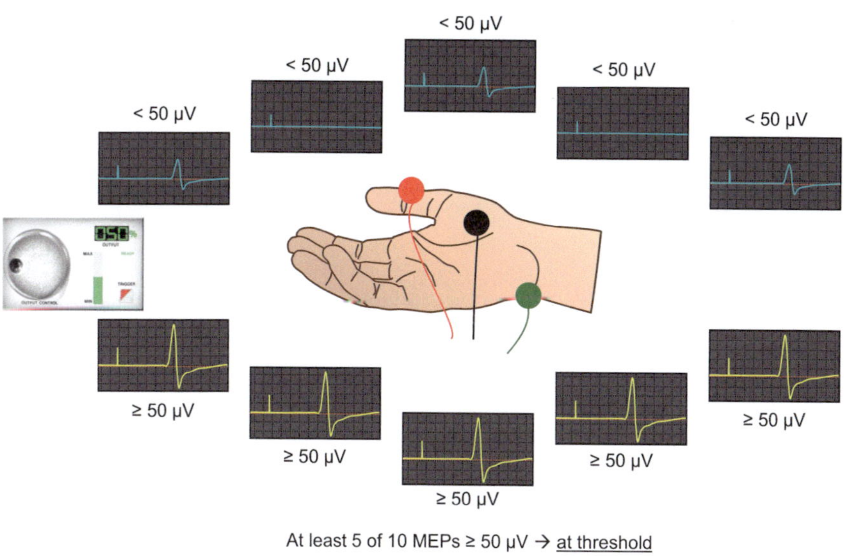

Step 4: Identify the location of the median nerve proximal to the wrist. It should be located along the midline between the flexor tendons. Apply a small amount of conductive gel (or saline) to the anode and cathode of the stimulating electrodes and place them against the skin, lengthwise on the wrist, with the cathode proximal. Ask the subject to relax his or her hand.

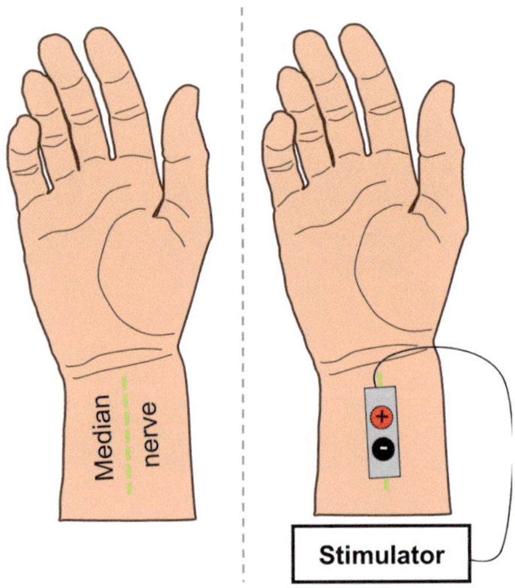

Step 5: Give pulses of electrical stimulation at increasing intensities until the "perceptual threshold" is reached. The perceptual threshold is defined as *the minimum stimulation intensity necessary for the subject to feel the electric stimulus*. The position of the stimulating electrodes may be adjusted as needed. Secure final position in place with medical tape.

Step 6: With TMS intensity set to 120% RMT (60% MSO in this example), deliver 10 TMS pulses (spaced every 6–10 s). The average MEPs elicited from these TMS pulses will serve as the baseline (pre-PAS) response.

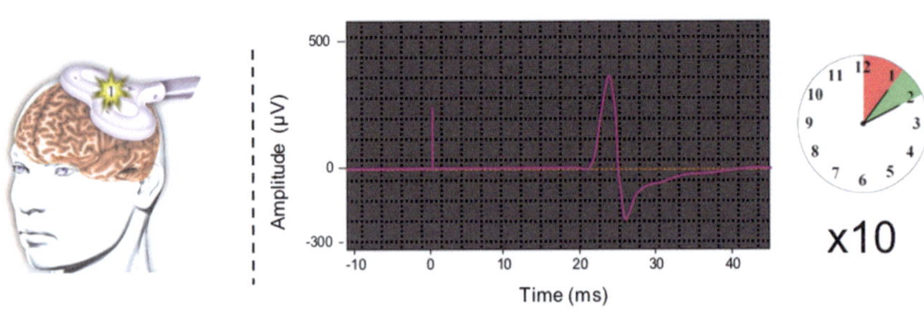

Step 7: Set the electrical stimulator to three times the perceptual threshold from Step 5.

Step 8: Set the ISI to 25 ms, which corresponds to the protocol for "facilitatory" PAS.

Step 9: Program the TMS and electrical stimulation devices to elicit 90 pairs of pulses every 20 s (0.05 Hz) for 30 min.

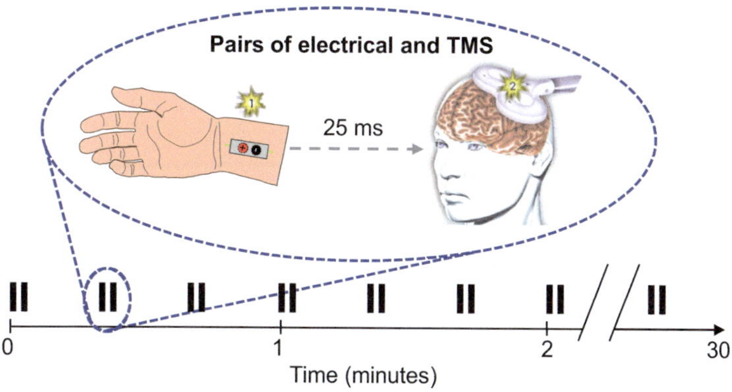

Step 10: After the train, deliver another 10 TMS pulses as in Step 6. The average MEPs elicited from these TMS pulses will serve as the post-PAS25 response.

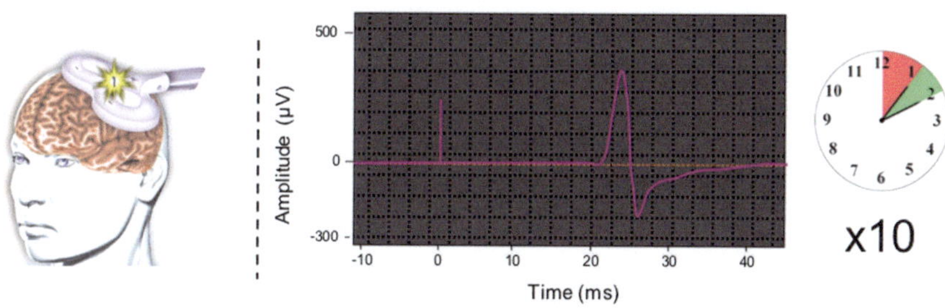

Notes:

- To perform PAS10, which corresponds to the protocol for "inhibitory" PAS, replace the ISI in Step 8 with 10 ms.
- Similar to SAI (*Sect. 5.3*), PAS can be individualized by setting the ISI based on the latency of the N20 somatosensory potential. The facilitatory effects of PAS are typically observed using ISIs that are approximately 2 ms following the N20 latency (N20+2 ms), while inhibitory effects of PAS are typically observed using ISIs that are approximately 5 ms before the N20 latency (N20-5 ms).
- PAS effects can be theoretically evaluated from most muscles in the body using this approach. However, the ISIs would need to be individually adjusted by recording an N20 EEG potential for the targeted muscle as described above.
- An alternative approach to using a set % RMT (i.e., 120%) for TMS is to use the intensity that elicits an average MEP of ~1 mV. This may improve the consistency of responses across individuals; however, a 1 mV MEP may be challenging to elicit in individuals with compromised integrity of the corticospinal tract. In

this case, a higher % RMT may be used (e.g., 130%), or alternatively, a smaller target MEP (~0.5–0.7) can be used.

- Alterative PAS approaches have been explored that pair the TMS pulse with a train of electrical stimuli instead of a single electrical stimulus. When trains of electrical stimuli are delivered to the periphery at 10 Hz, the temporal window for the conditioning effects of the electrical stimuli on the TMS pulses is expanded from 8 ms (for a single electrical stimulus) to potentially hundreds of ms depending on the duration of the electrical pulse train. Importantly, it does not appear to matter whether the TMS pulse is delivered after the first or last electrical stimulus of the train.
- If your TMS machine is not capable of delivering a train of stimuli, you can perform Step 9 by manually triggering each pair of electrical and TMS pulses. Each pair should be spaced 20 s apart and repeated for a total of 90 times (30 min). Note that you will still need a timing mechanism to connect the electrical stimulator to the TMS machine to deliver the proper ISI within each pair.
- There is an alternative approach to PAS that seeks to have the electrical and TMS pulses interact at the level of the spinal cord rather than in the cortex. This is known as spinal associative stimulation (SAS) and can be achieved by delivering the TMS pulse *prior to*, rather than after the peripheral electrical stimulus. For example, in the lower limbs, a TMS pulse at 80% AMT (*Sect. 3.5*) is delivered ~20 ms prior to applying an electrical stimulus to the posterior tibial nerve, with the pair repeated every 10 ms (0.1 Hz) for 15 min (replacing Steps 8 and 9 above). This approach shows the potential to improve recovery of function following spinal cord injury. A schematic is shown below.

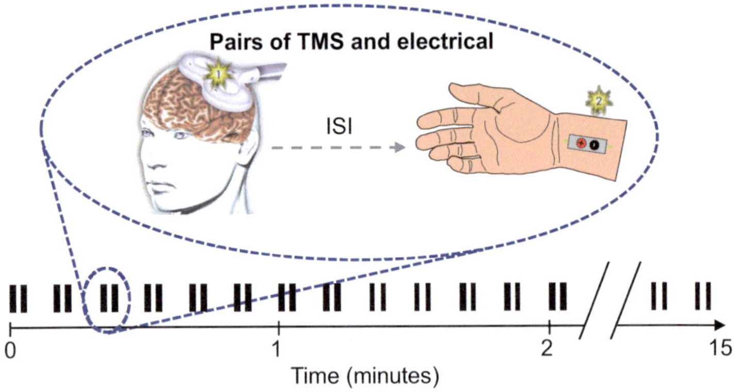

6.3 Repetitive TMS as Clinical Treatment for Major Depressive Disorder

As discussed in *Sect. 6.1*, rTMS protocols can be used to modulate cortical activity. While this change typically lasts only a short while after the end of stimulation, studies have shown that repeating the same rTMS protocol (e.g., daily) can lead to a more lasting change. This approach forms the basis of many therapeutic applications of rTMS. In the mid-1990s, several proof-of-principle studies were conducted in patients with major depressive disorder (MDD) showing that high-frequency rTMS to the left DLPFC led to a reduction in depressive symptoms. This paved the way for eventual FDA clearance in 2008 of the NeuroStar™ TMS system, the first in a growing number of TMS systems to treat medication-resistant MDD with a specific protocol (*Sect. 1.3*). Since then, the use of rTMS for depression has expanded globally with many new TMS systems receiving regulatory clearance and new protocols being tested, both for MDD and an increasing number of neuropsychiatric conditions. Despite its growing prevalence, rTMS is still not a common first-line treatment but is primarily indicated for patients who have failed at least one pharmacological treatment. In this population of patients with medication-resistant refractory MDD, approximately 60% respond to rTMS therapy. The identification of markers that can be used to predict treatment efficacy and optimize treatment parameters are areas of active research.

The original protocol cleared by the FDA and still widely used consists of 4-s trains of 10 Hz rTMS with a 26-s inter-train interval that is repeated 75 times for a total of 3000 pulses over 37.5 min. Intensity is typically set at 120% RMT (*Sect. 3.4*) as determined using visible twitch method (based on the expectation that most practitioners would not necessarily have access to EMG systems to record MEPs). This pattern is repeated daily (typically except for weekends) for 4 weeks (with an additional 2 weeks if needed).

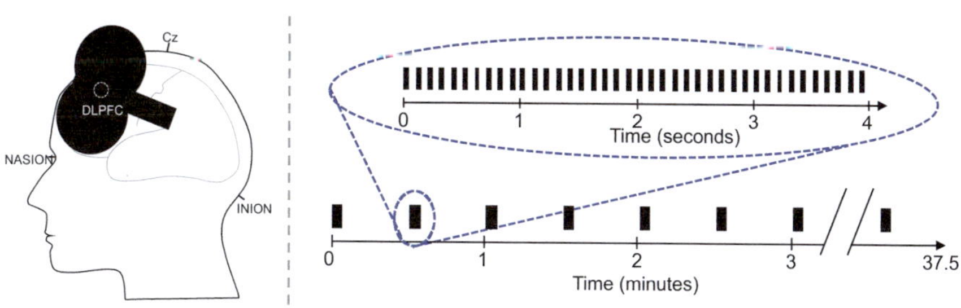

There are a number of approaches to locating the stimulation site within the left DLPFC. These include centering the TMS coil over coordinate F3 of the International "10-20 System" for EEG electrode placement (*Sect. 3.2*) or at a set distance measured from the motor hotspot (*Sect. 3.3*). While scalp-based targeting remains widespread, there is increasing utilization of MRI-guided frameless stereotaxic "neuronavigation" systems that allow targeting of TMS to specific brain areas. While neuronavigation can improve accuracy, there are additional factors such as cost and availability, and there is still no consensus as to which brain area within the DLPFC is the optimal site to target. Recent neuroimaging-based studies have suggested that the optimal DLPFC target for depression may be based on structural or functional connectivity with other brain areas.

The mechanisms underlying the anti-depressant response to rTMS are not completely understood but may relate to modulation of the DLPFC and related brain regions. rTMS efficacy in terms of symptom relief has been likened to electroconvulsive therapy, though, importantly, without inducing seizures.

Below, we will demonstrate the standard clinical 10 Hz rTMS protocol to the left DLPFC using a generic stimulator and figure-8-coil. Different TMS devices (*Sect. 1.3*) may have specific procedures or alternative steps and you should consult with the manufacture before use.

Note As the pulse trains are generated in rapid succession through a single coil, this procedure requires a TMS machine with multiple charging units and an air- or fluid-based cooling system to prevent the coil from overheating.

6.3.1 Steps to Administer the Standard 10 Hz rTMS Protocol to the Left DLPFC for the Treatment of MDD

Step 1: Perform a brief check-in with the patient to assess mood and address any safety concerns (medication changes, alcohol consumption, medical procedures, etc.) prior to administering treatment.

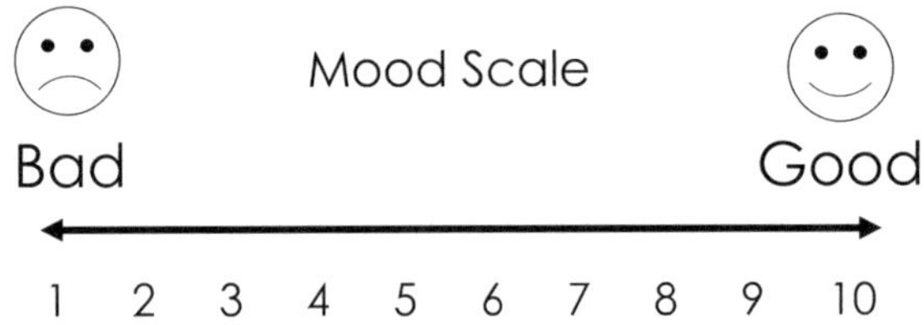

Step 2: Ensure that the patient is comfortable in the treatment chair, providing pillows or cushions as requested.

Step 3: Determine the APB motor hotspot (*Sect. 3.3*) using visible twitch as the motor response. Remember that when determining the hotspot, one should start at a low intensity (i.e., below threshold) and increase until responses are large and consistently observed (i.e., above threshold). The hands should be completely at rest while finding the hotspot.

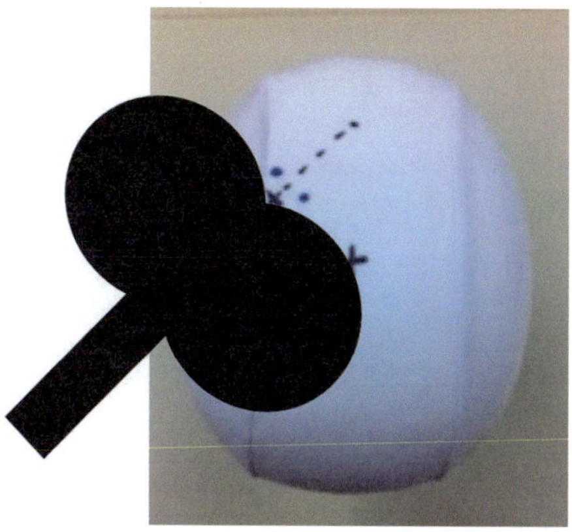

Step 4: Determine the RMT for the APB hotspot using the visible twitch method (*Sect. 3.4*). For this example, we will assume an RMT of 50% MSO.

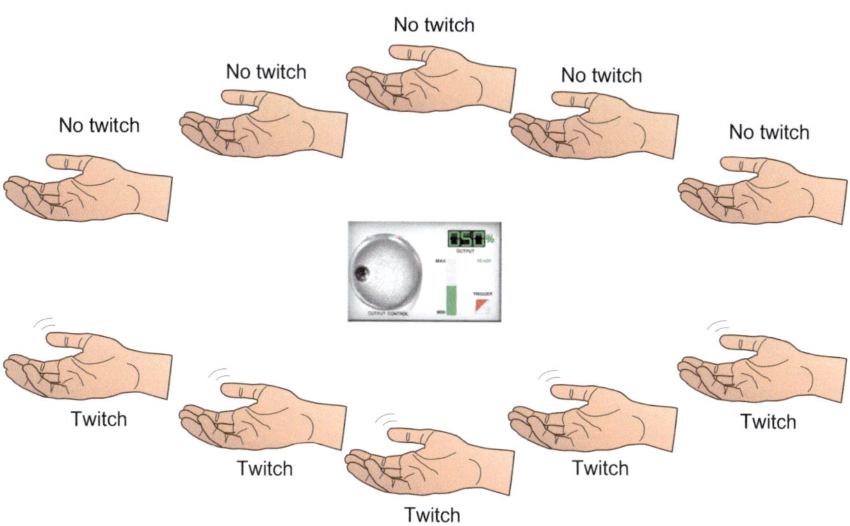

Step 5: Locate the DLPFC by measuring 5.5 cm directly anterior from the motor hotspot (parallel to the mid-sagittal plane). Make a mark on the cap to identify this spot.

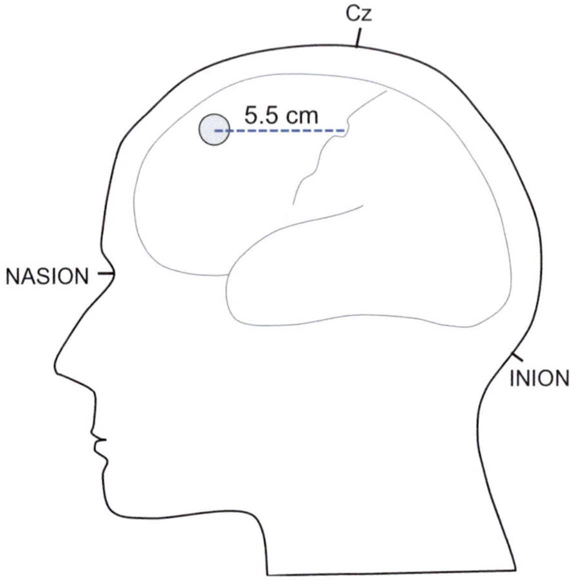

Step 6: Set the device parameters to deliver 3000 pulses at 10 Hz, with a 4-s—on, 26-s—off pattern.

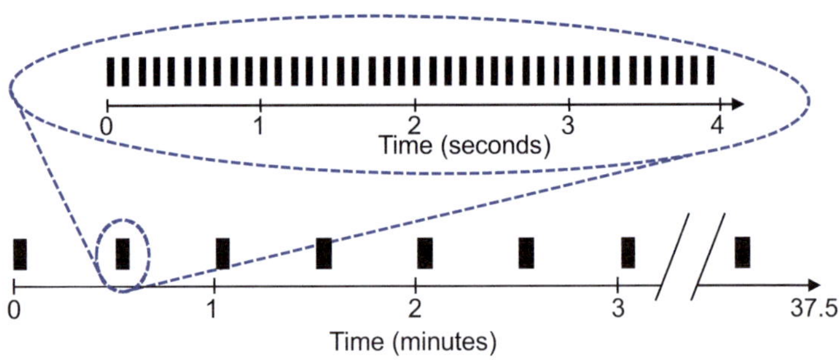

Step 7: Set intensity to 120% RMT from Step 4 (60% MSO in this example).

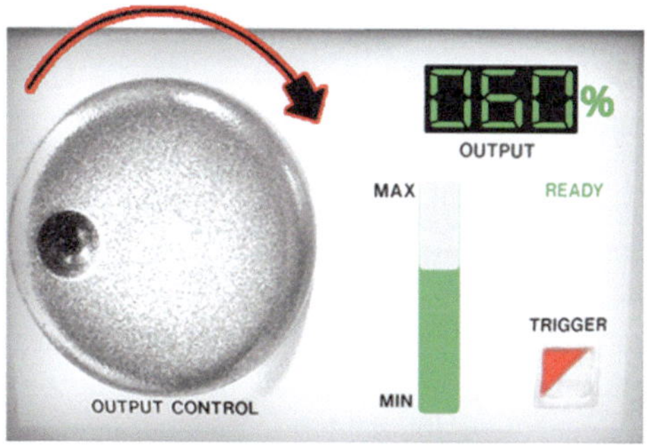

Step 8: Place the center of the coil over the DLPFC site from Step 5 and orient the coil so that the handle is ~45° from the midline—similar to the orientation for the M1 hotspot.

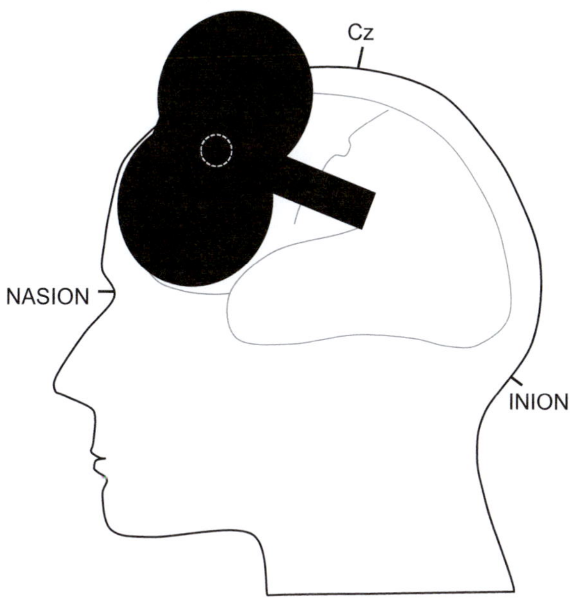

Step 9: Deliver the rTMS Train

Notes:

- It is important that everyone in the treatment room, including the patient, the technician, and any observers or family members of the patient, wear earplugs during the entire treatment. This is good practice for any TMS procedure, but it is particularly important for clinical treatment given the duration and number of visits.

- The FDA-cleared[1] 10 Hz protocol described above is currently the most used and has the strongest evidence supporting its efficacy. However, there are other protocols that are in use or being tested. Some of these, such as iTBS (*Sect. 6.1*), have received FDA clearance, while others, such as 1 Hz to the right DLPFC (alone or in conjunction with 10 Hz to the left DLPFC), are being used "off-label." Given the wide range of protocols with limited supporting data on their efficacy, and the rapidly changing regulatory landscape, it is outside the scope of this manual to cover all potential alternatives.
- As an individual's cortical excitability may change due to several different factors (e.g., changes in medication or amount of sleep), it is a good idea to recheck RMT between sessions and adjust treatment intensity accordingly.
- There is an ongoing debate in the field as to how to target the DLPFC to achieve the greatest clinical response. There is evidence that the 5 cm (from the motor hotspot) measure in the original FDA-cleared protocol may not correspond to the DLPFC in some patients and stimulating a scalp site that is slightly more anterior and lateral may lead to a better clinical response. A number of protocols have been proposed to improve the likelihood of reaching the DLPFC. These include (1) measuring 6 cm anterior + 1 cm lateral from the motor hotspot (which is an improvement over the 5 cm rule, but still suffers from the limitation of using the same measurement in individuals with different head sizes), (2) using an EEG coordinate such as F3 (which accounts for head size and is easy to implement with simplified algorithms such as the "Beam-F3" method), or (3) using MRI-guided neuronavigation (which is more accurate but also more costly and requires a subject's MRI to achieve superiority over scalp-based targeting).
- When using scalp-based targets (e.g., F3 or a fixed distance from the M1 hotspot), the coil can be positioned over DLPFC using one of two references: either with the center of the coil over the site itself or with a mark representing where the front of the coil should be positioned (*Sect. 3.3*). If using the front of the coil approach to aid in its placement, note the distance between the front of the coil and its center and adjust any marks to account for this distance.
- During treatment, patients may talk with the technician, read, use an electronic device, or relax. However, patients should remain awake throughout the treatment. If the patient appears to be falling asleep, you can wake them up by speaking to them or gently tapping them on the shoulder.
- Patients may take sips of water in between rTMS trains. However, they should not eat, chew gum, or have anything else in their mouth during the treatment.
- Periodically check in with the patient during the treatment. If the treatment is too uncomfortable, the intensity may be adjusted to a lower % RMT and slowly increased over the course of one or more sessions until the patient is accustomed to the treatment intensity, and can tolerate stimulation at 120% RMT.

[1] *Please note: while we refer to "protocol" in this chapter, FDA clearance is not for the stimulation pattern but for specific devices that use the mentioned protocol.*

– Check in with the patient after the treatment and document any side effects or adverse events (*Sect. 2.1*). This is good practice for any TMS procedure, but it is particularly important for clinical treatments.

Further Reading

Section 6.1

Huang Y-Z, Chen R-S, Rothwell JC, Wen H-Y. The after-effect of human theta burst stimulation is NMDA receptor dependent. Clin Neurophysiol. 2007 May;118(5):1028–32.
Huang, Y. Z., Edwards, M. J., Rounis, E., Bhatia, K. P., & Rothwell, J. C. (2005). Theta burst stimulation of the human motor cortex. Neuron, 45(2), 201-206.
Maeda F, Keenan JP, Tormos JM, Topka H, Pascual-Leone A. Modulation of corticospinal excitability by repetitive transcranial magnetic stimulation. Clin Neurophysiol. 2000 May;111(5):800–5.
Maeda F, Keenan JP, Tormos JM, Topka H, Pascual-Leone A. Interindividual variability of the modulatory effects of repetitive transcranial magnetic stimulation on cortical excitability. Exp Brain Res. 2000 Aug;133(4):425–30.
Suppa A, Huang YZ, Funke K, Ridding MC, Cheeran B, Di Lazzaro V, Ziemann U, Rothwell JC. Ten Years of Theta Burst Stimulation in Humans: Established Knowledge, Unknowns and Prospects. Brain Stimulation [Internet]. 2016 [cited 2016 Feb 1];

Section 6.2

Carson, R. G., & Kennedy, N. C. (2013). Modulation of human corticospinal excitability by paired associative stimulation. *Frontiers in Human Neuroscience, 7*.
Cortes M, Thickbroom GW, Valls-Sole J, Pascual-Leone A, Edwards DJ. Spinal associative stimulation: a non-invasive stimulation paradigm to modulate spinal excitability. Clin Neurophysiol. 2011 Nov;122(11):2254-9. https://doi.org/10.1016/j.clinph.2011.02.038. Epub 2011 Apr 23. PMID: 21524606; PMCID: PMC3538079.
Müller-Dahlhaus F, Ziemann U, Classen J. Plasticity resembling spike-timing dependent synaptic plasticity: the evidence in human cortex. Front Synaptic Neurosci. 2010 Jul 30;2:34. https://doi.org/10.3389/fnsyn.2010.00034. eCollection 2010.
Stefan, K., Kunesch, E., Benecke, R., Cohen, L. G., & Classen, J. (2002). Mechanisms of enhancement of human motor cortex excitability induced by interventional paired associative stimulation. *The Journal of Physiology, 543*(2), 699-708.
Stefan, K., Kunesch, E., Cohen, L. G., Benecke, R., & Classen, J. (2000). Induction of plasticity in the human motor cortex by paired associative stimulation. *Brain, 123*(3), 572-584.
Suppa A, Quartarone A, Siebner H, Chen R, Di Lazzaro V, Del Giudice P, Paulus W, Rothwell JC, Ziemann U, Classen J. The associative brain at work: Evidence from paired associative stimulation studies in humans. Clin Neurophysiol. 2017 Nov;128(11):2140-2164. https://doi.org/10.1016/j.clinph.2017.08.003. Epub 2017 Sep 5.
Wischnewski M, Schutter DJ. Efficacy and time course of paired associative stimulation in cortical plasticity: Implications for neuropsychiatry. Clin Neurophysiol. 2015 May 8. pii: S1388-2457(15)00312-0. https://doi.org/10.1016/j.clinph.2015.04.072.
Wolters A, Schmidt A, Schramm A, Zeller D, Naumann M, Kunesch E, Benecke R, Reiners K, Classen J. Timing-dependent plasticity in human primary somatosensory cortex. J Physiol.

2005 Jun 15;565(Pt 3):1039-52. https://doi.org/10.1113/jphysiol.2005.084954. Epub 2005 Apr 21. PMID: 15845584; PMCID: PMC1464551.

Section 6.3

Beam W, Borckardt JJ, Reeves ST, George MS. An efficient and accurate new method for locating the F3 position for prefrontal TMS applications. Brain Stimul. 2009 Jan 1;2(1):50–4.

Fox MD, Buckner RL, White MP, Greicius MD, Pascual-Leone A. Efficacy of transcranial magnetic stimulation targets for depression is related to intrinsic functional connectivity with the subgenual cingulate. Biol Psychiatry. 2012 Oct 1;72(7):595–603.

George M.S., Lisanby S.H., Avery D., McDonald W.M., Durkalski V., Pavlicova M., Anderson B., Nahas Z., Bulow P., Zarkowski P., et al. Daily Left Prefrontal Transcranial Magnetic Stimulation Therapy for Major Depressive Disorder: A Sham-Controlled Randomized Trial. Arch. Gen. Psychiatry. 2010;67:507–516.

George MS, Wassermann EM, Williams WA, Callahan A, Ketter TA, Basser P, Hallett M, Post RM. Daily repetitive transcranial magnetic stimulation (rTMS) improves mood in depression. Neuroreport. 1995 Oct 2;6(14):1853-6. https://doi.org/10.1097/00001756-199510020-00008.

Lefaucheur JP, Aleman A, Baeken C, Benninger DH, Brunelin J, Di Lazzaro V, Filipović SR, Grefkes C, Hasan A, Hummel FC, Jääskeläinen SK, Langguth B, Leocani L, Londero A, Nardone R, Nguyen JP, Nyffeler T, Oliveira-Maia AJ, Oliviero A, Padberg F, Palm U, Paulus W, Poulet E, Quartarone A, Rachid F, Rektorová I, Rossi S, Sahlsten H, Schecklmann M, Szekely D, Ziemann U. Evidence-based guidelines on the therapeutic use of repetitive transcranial magnetic stimulation (rTMS): An update (2014-2018). Clin Neurophysiol. 2020 Feb;131(2):474-528. https://doi.org/10.1016/j.clinph.2019.11.002. Epub 2020 Jan 1. Erratum in: Clin Neurophysiol. 2020 May;131(5):1168-1169.

Pascual-Leone A, Rubio B, Pallardó F, Catalá MD. Rapid-rate transcranial magnetic stimulation of left dorsolateral prefrontal cortex in drug-resistant depression. Lancet. 1996 Jul 27;348(9022):233-7. https://doi.org/10.1016/s0140-6736(96)01219-6.

Perera T, George MS, Grammer G, Janicak PG, Pascual-Leone A, Wirecki TS. The Clinical TMS Society Consensus Review and Treatment Recommendations for TMS Therapy for Major Depressive Disorder. Brain Stimul. 2016 May-Jun;9(3):336-346. https://doi.org/10.1016/j.brs.2016.03.010. Epub 2016 Mar 16. PMID: 27090022; PMCID: PMC5612370.

Glossary

Transcranial magnetic stimulation (TMS) A noninvasive method of electrical stimulation by electromagnetic induction.

Motor cortex (M1) The region of the brain that sends output to control voluntary muscles; also referred to as Brodmann area 4.

Motor-evoked potential (MEP) A compound muscle response elicited by TMS and recorded by electromyography (EMG).

Latency The time (in ms) between the TMS pulse and the start (or first peak) of the MEP.

Amplitude A measure (in mV) of the magnitude of the MEP, calculated as the peak-to-peak difference (maximum–minimum).

Central motor conduction time (CMCT) The propagation time of TMS along the first motor neuron, calculated as the difference between the total conduction time and the peripheral conduction time.

Total conduction time (TCT) The MEP latency.

Peripheral conduction time (PCT) The time of TMS or an electrical stimulus to travel along the second motor neuron from spinal cord/roots to muscle.

F-wave method An alternative to measuring the peripheral conduction time using electrical stimulation of a peripheral motor nerve in both orthodromic (M-wave) and antidromic (F-wave) directions.

Repetitive TMS (rTMS) The application of TMS as a patterned train to modify or disrupt cortical activity.

High-frequency rTMS An on–off pattern of rTMS, typically at a frequency of >5 Hz; thought to lead to increased neural activity.

Low-frequency rTMS A pattern of continuous rTMS, typically at a frequency of 1 Hz; thought to lead to decreased neural activity.

Theta-burst stimulation A complex pattern of rTMS consisting of ultra-high frequency bursts of pulses (usually 50 Hz) repeated every ~5 Hz.

iTBS A pattern of TBS consisting of 2-s trains with an 8-s break; thought to lead to increased neural activity.

D. J. Edwards et al., *A Practical Manual for Transcranial Magnetic Stimulation*,
https://doi.org/10.1007/978-3-031-62304-2

cTBS A continuous pattern of TBS; thought to lead to decreased neural activity.

Long-term depression (LTD) A type of neural plasticity characterized by a prolonged decrease in synaptic firing.

Long-term potentiation (LTP) A type of neural plasticity characterized by a prolonged increase in synaptic firing.

Electromagnetic induction The generation of an electrical current by a varying magnetic field.

Electric field (e-field) The area in neural tissue surrounding an induced electrical current.

Magnetic field The area of magnetic influence within which an electrical current can be generated.

Inducing (primary) current The electrical current passed through a TMS coil used to produce a varying magnetic field.

Induced (secondary) current The electrical current induced in neural tissue within a varying magnetic field.

Major depressive disorder A clinical disorder characterized by a prolonged pervasive reduction in mood.

FDA clearance The formal process in the USA to approve the use of a TMS or other device with a protocol for a specific indication.

Neuromodulation The use of external stimulation to induce a lasting change brain activity.

Neurostimulation The use of external stimulation to elicit an individual neural response, e.g., an MEP.

TMS system The main device used to store and shape current and discharge it through a coil.

Capacitor The part of the TMS system that stores electrical energy as an electric field.

Resistor The part of the TMS system that shapes or otherwise changes the voltage.

Thyristor The part of the TMS system that acts as a switch to allow the rapid release of stored electrical energy.

TMS coil The part of the TMS system that is in contact with the participant and generates a magnetic field when current is passed through it. It comprises loops of copper wire encased in hard plastic.

Round coil The original TMS coil design in the shape of a circle that produces a consistent magnetic field strength all the way around.

Figure-of-8 coil A "focal" TMS coil consisting of two circles that produces a summated, or stronger, magnetic field at the junction of the circles. Also known as a "butterfly coil."

TMS pulse The rapidly varying electromagnetic field induced by passing a brief but strong current through the TMS coil; the electromagnetic unit of TMS.

Monophasic pulse A TMS pulse defined by an induced current in single direction (e.g., posterior-to-anterior, or PA) relative to the targeted brain area.

Biphasic pulse A TMS pulse defined by an induced current in two successive opposing directions (e.g., anterior-to-posterior—posterior-to-anterior, or AP-PA) relative to the targeted brain area.

Corticospinal volley The output of TMS of the motor cortex measured along the corticospinal tract.

Direct (D)-wave A corticospinal volley with the shortest latency, likely resulting from direct stimulation of the pyramidal neuron.

Indirect (I)-waves A series of corticospinal volleys with successively longer latencies, likely resulting from indirect stimulation of the pyramidal neuron through interneurons.

Cortical mapping The systematic recording of neurophysiologic (e.g., MEPs), perceptual (e.g., phosphenes), or behavioral (e.g., language suppression) response from stimulation of many sites within a particular region of cortex.

Center of gravity (CoG) The average of multiple x, y, z coordinates on MRI; used to describe the responses of cortical mapping.

Maximum stimulator output (MSO) The greatest intensity (primary electrical field) that a given TMS system can produce. This value is unique to each TMS system and the range of intensities are measured from 0 to 100 in %MSO units.

Resting motor threshold (RMT) The minimum intensity of TMS that elicits a minimal MEP (e.g., 50 μV) or visible muscle twitch at least 50% of the time with the target muscle at rest.

Active motor threshold (AMT) The minimum intensity of TMS that elicits a moderate MEP (e.g., 200 μV) at least 50% of the time during tonic contraction of the target muscle.

Phosphene threshold The minimum intensity of TMS that elicits an unambiguous perception of a phosphene at least 50% of the time.

Inter-train interval The time between successive trains of rTMS.

Cortical/corticospinal reactivity The response of the brain to a stimulus such as TMS.

Cortical/corticospinal excitability A measurement of the response of the brain to a stimulus such as TMS (e.g., RMT or average MEP amplitude).

Suprathreshold An intensity of TMS above a given threshold (e.g., resting motor threshold).

Subthreshold An intensity of TMS below a given threshold (e.g., resting motor threshold).

Motor hotspot The location in the primary motor cortex where TMS elicits the largest and most consistent motor response (MEP or visible twitch) for a given muscle.

Phosphene hotspot The location in a region of the visual cortex where TMS elicits the most vivid and consistent perception of a phosphene.

Intrinsic hand muscles Four muscle groups (thenar, hypothenar, interossei, and lumbrical muscles) that contain the origin and insertions within the carpal and metacarpal bones.

Thenar One of the muscle groups of the intrinsic hand muscles located at the base of the thumb that includes the abductor pollicis brevis, adductor pollicis brevis, opponens pollicis brevis, and flexor pollicis brevis.

Abductor pollicis brevis (APB) An intrinsic hand muscle and one of the muscles of the thenar that abducts the thumb (moves it away from the body).

First dorsal interosseous (FDI) An intrinsic hand muscle located between the thumb and index finger that abducts the index finger (moves it away from the body).

Abductor digiti minimi (ADM) An intrinsic hand muscle located at the outside of the pinky finger that abducts the pinky finger (moves it away from the body).

Cranial vertex/Cz The top-center of the scalp located at the intersection of the mid-sagittal (nasion-to-inion) and inter-aural (between left and right preauricular areas) lines.

Nasion A skull landmark at depression on the bridge of the nose between the eyes.

Inion A skull landmark characterized by a bony ridge on the occipital bone at the back of the scalp.

Preauricular area A skull landmark located anterior and slightly superior to the external opening of the ear.

Electromyography (EMG) The measurement of electrical activity in a muscle using electrodes.

Volt A unit of electrical potential.

Sampling rate The resolution of a recording defined as the number of samples per unit time.

Stimulus artifact A non-physiologic signal in an electromyographic or other electrical recording caused by an electrical or electromagnetic stimulator.

TMS pulse artifact A stimulus artifact caused by TMS.

Silent period (SP) A neurophysiologic phenomenon elicited by a TMS of the motor cortex under voluntary contraction and characterized by the GABAergic-mediated suppression of electromyographic activity despite continued voluntary contraction of the target muscle.

Contralateral silent period (cSP) A silent period recorded in a muscle on the opposite side as the TMS target.

Ipsilateral silent period (iSP) A silent period recorded in a muscle on the same side as the TMS target.

Input–output (IO) curve The graphical, often sigmoidal or "s"-shaped, relationship between a stimulus parameter (e.g., TMS intensity) and an outcome measure (e.g., MEP amplitude). Also known as a stimulus–response curve or a recruitment curve.

Intracortical inhibition/facilitation The decrease/increase of an MEP amplitude or similar TMS outcome measure by the conditioning effects of a prior TMS pulse believed to be mediated by GABA and glutamate, respectively.

Conditioning stimulus (CS)/test stimulus (TS) In a paired-pulse paradigm, the conditioning stimulus is the one that proceeds and exerts an effect on the test stimulus.

Interstimulus interval (ISI) The time between two stimuli, such as two pulses in a paired-pulse TMS protocol or two trials of single-pulse TMS.

Paired-pulse stimulation A TMS paradigm involving two TMS pulses, a conditioning stimulus and a test stimulus, separated by a short delay.

Long-interval intracortical inhibition (LICI) A paired-pulse TMS protocol involving two suprathreshold TMS pulses to the primary motor cortex spaced ~50–200 ms apart.

Short-interval intracortical inhibition (SICI) A paired-pulse TMS protocol involving a subthreshold TMS followed ~1–6 ms by a suprathreshold TMS pulse to the primary motor cortex.

Intracortical facilitation (ICF) A paired-pulse TMS protocol involving a subthreshold TMS followed ~8–25 ms by a suprathreshold TMS pulse.

Cerebellar-brain inhibition (CBI) A paired-pulse TMS protocol involving a suprathreshold TMS pulse to the cerebellum followed ~5–10 ms by a suprathreshold TMS pulse to the primary motor cortex.

Interhemispheric inhibition (IHI) A paired-pulse TMS protocol involving a suprathreshold TMS pulse to the primary motor cortex in one hemisphere followed ~5–30 ms by a suprathreshold TMS pulse to the primary motor cortex in the opposite hemisphere.

Short-afferent Inhibition (SAI) A TMS protocol pairing an electrical stimulus to a peripheral nerve followed ~21–25 ms by a suprathreshold TMS pulse to the primary motor cortex on the opposite side.

Triple-pulse stimulation technique (TST) A TMS protocol combining two peripheral electrical stimuli with a suprathreshold pulse to the primary motor cortex at specific timings to measure axonal loss.

Paired-associative stimulation (PAS) A TMS protocol involving repeated pairing of a peripheral electrical stimulus with a TMS pulse to the primary motor cortex to change cortical excitability.

MIX
Papier aus verantwortungsvollen Quellen
Paper from responsible sources
FSC® C105338

If you have any concerns about our products,
you can contact us on
ProductSafety@springernature.com

In case Publisher is established outside the EU,
the EU authorized representative is:
Springer Nature Customer Service Center GmbH
Europaplatz 3, 69115 Heidelberg, Germany

Printed by Libri Plureos GmbH
in Hamburg, Germany